Application of Botany in Horticulture

Application of Botany in Horticulture

André Génin
Horticultural Engineer, E.N.S.H.

United Book Prints

First Reprint in India
United Book Prints
4806/24, Ansari Road, Daryaganj
New Delhi - 110 002
Tel.: +91-11-41511055
E-mail: unitedbookprints@gmail.com

Print, 2021

ISBN: 978-81-926860-5-9
Reprint : 2014

Reprint in India

Acknowledgements

Our special thanks to the organisations that provided us with scientific and technical information.

• *Messrs Lajoux and Morel* of the National Interprofessional Horticultural Committee at the Angers Station (for the material on *growth regulators*).

• *Madame Mulocher* of the Plant Physiology Research Laboratory at Angers (for the material on in vitro *multiplications*).

Author's note

• *On the nomenclature of plants referred to in this work* In all cases where we have had to provide an example, in other words, a particular plant, we have referred to a horticultural plant indicated in French (here translated in English—Ed.) or in Latin.

We have complied with the rules laid down in the International Code of Botanical Nomenclature.

We now expect the reader to be aware of the manner of writing explained in *Nomenclature* (§ 1.1.3, Chapter 8, p. 178).

In this way we will avoid confusion between the names of genera, species, varieties or cultivars, that should or may be written, depending on the case, with an initial capital or small letter.

• *On horticultural applications*
All horticultural applications shown in the margins are indicated in the text by an asterisk.

Contents

Chapter 1

Making a Herbarium

Chapter 2

The Morphology of Plants

Chapter 5

Plant Physiology

Chapter 6

Plant Reproduction

Chapter 7

Genetics

Chapter 8

Systematics

Chapter 9

Plant Geography and Ecology

Chapter 1

Making a Herbarium

1
The Plant World

1.1 Botany is an observational science

It is not an abstract science.
It calls for a knowledge of natural materials that are studied and which are: *plants.*
Indeed, how can we talk about engines if we have never seen one being operated?
How can we talk about paintings if we have never been to an exhibition or an art gallery?
It is, therefore, essential to know about the materials that are studied in botany.

1.2 It is necessary to familiarise oneself with the materials of the plant world

For this, it is essential:
to handle them,
to observe them carefully,
to compare them, and
to classify them.

1.3 Why should we classify these materials?

This is because man ardently desires to give a name to each object and, above all, to compare it with others that are already known to him.
Though this feeling is based on a selective organisation it must be borne in mind that, for a beginner, it will create some difficulty and alarm because of the thousands, or even tens of thousands, of different plants.

How shall we cross these initial obstacles?
As a first step, we recommend the making of a herbarium.

2 Herbarium

2.1 *What is a herbarium?*

A herbarium maintains a collection of dried plants, set on a sheet of paper or parchment that, apart from the physical aspect, also provides the following information:

— the name of the family to which the plant belongs;

— the name of the plant (genus, species and variety) in French (or any other language) and Latin;

— the place where it was collected and the date;

— an ecological description (woodland, meadow, cultivated field, etc.) of the place where it was found;

— the known properties of the collected plant.

It is clear that the main difficulty is with respect to compiling this data card.
Let us begin by collecting plants.

2.2 *Collecting plants*

It is essential to preserve only those plants that are as "complete" as possible, in other words, those that have all their vegetative parts.

— Root (not always easy to ensure);

— Stem, leaf, but it is absolutely necessary to preserve the flowers and, if possible, the fruits (*see chapter 8, "Systematics"*).

Consequently, "collection" is not a one-time affair but should continue for at least nine months throughout the year.

2.3 *Identification of a plant*

This cannot be done without a guide, at least at the beginning, or without sufficient experience.

We also need to have a copy of "La Flore" (Flora), a fairly thick book, and a somewhat complicated one, that will help us as a guide to find the names of the plants which we have collected.
It is essential, after this, to warn the beginner that identification of plants is somewhat complex and the final result is often disappointing. ...
Only successive attempts and considerable perseverance will make the student familiar with the manual.
Nevertheless, we must begin somewhere. Let us, therefore, attempt to do so with:
La Flore (Flora) by G. Bonnier
It will indicate various methods to achieve positive results.

2.3.1 A herb: shepherd's purse (*Capsella bursa-pastoris*): Cruciferae..
We successively pass over the numbers 1, 2, 3, 4, 506, 507, 508, 509, 510, 512, 513, 523, 526, 527, 528, 529(b) and find the name we are looking for.

2.3.2 A climbing shrub: the fruiting bramble (*Rubus fructicosus*): Rosaceae.
We find it while passing over the following numbers: 1, 2, 3, 4, 5, 6, 88, 107, 112, 113, 118, 119, 120, 121(b).

2.3.3 A tree: the chestnut (*Castanea vulgaris*): Cupuliferae
We find the name after going over the numbers 1, 2, 942, 961, 1044, 1047, 1054, 1057, 1058, 1059, 1060(a).

2.3.4 Conclusions[1]

We do not insist that Bonnier's Flora is the one to be used. There are other manuals.
What is essential is to know how to use the one we have and, above all, to acquaint everyone with the difficulties that arise.
One must not be in a hurry; each step should be taken only when we are sure of the earlier result.
Whatever the case, a study of Chapter 8, *"Systematics"*, will help to identify plants.
Nevertheless, attempts must be made and the results tested because it is often difficult to identify a plant that has been dry for months and even more so when it was dried several years earlier.

1 Recognition of the spontaneous vegetation in an area enables us to know its essential properties because each type of soil has its own flora.

2.4 *Preservation of plants*

While collecting, we are, therefore, aware that all the organs must be present and undamaged.
This is a delicate task and nothing should be spoiled by being in a hurry, thereby leaving behind a root, a stem, or several leaves.
Collect some more specimens.
Next, nothing should be allowed to fade and the only solution is to treat the plants after collection and as quickly as possible. For this:

— Spread the organs, flat, between sheets of paper (newspaper or, better still, blotting paper).
— Place the leaves so treated in a special press or between thick books.
— When the plants are well dried we need only to mount them on "herbarium sheets".

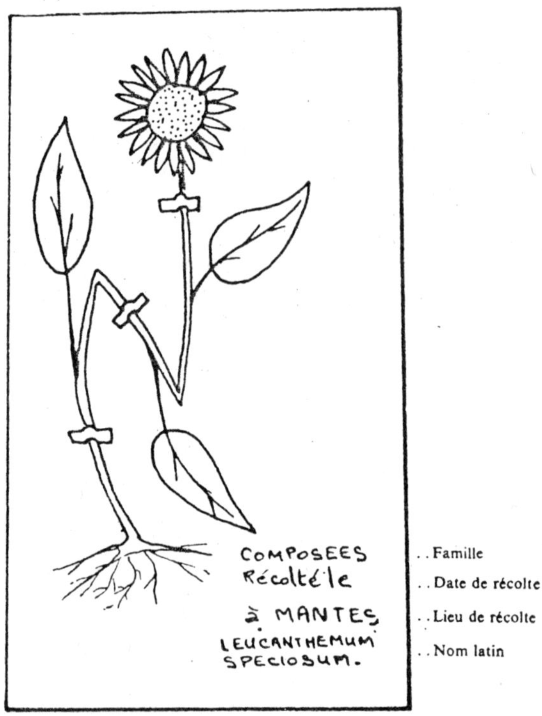

Fig. 1.1. Prepared herbarium sheet

2.5 Preparation of a herbarium sheet

(see Fig. 1.1)

We arrange all the collected organs and spread them out as far as possible, fixing them with cellotape in order to enable easy observation and comparison.

Note:

— Dried plants are very brittle.

— Do not classify plants that are not completely dry: they will become overgrown with mould.

Chapter 2

The Morphology of Plants

1
Morphology of Flowering Plants

1.1 Description

We collected various herbaceous plants to make our herbarium.
We also observed other shrubs as well as trees.
Let us recall what we observed:

1.1.1 A tall apple tree from the orchards of Normandy (*Malus communis*) that was uprooted in September, with:
a) A system of main ramifications in the soil; these were the primary, secondary and tertiary roots.
b) A trunk of about 2 m covered with a somewhat rough bark and aerial ramifications, the branches: primary, secondary and some tertiary, fairly thick and long.
On these smaller branches we observed:

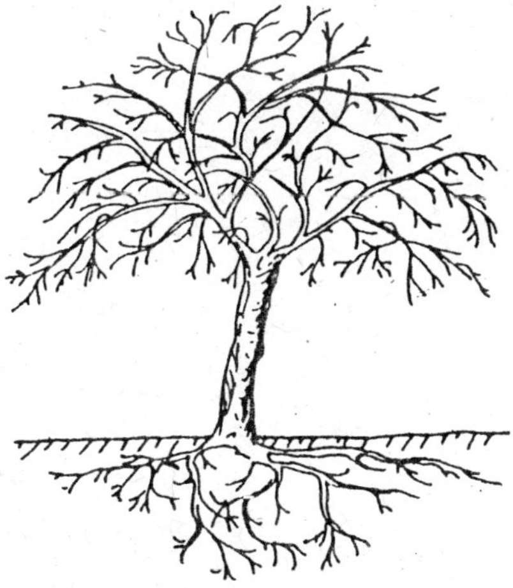

Fig. 2.1. Schema of a tall apple tree in the winter, with its aerial branches, trunk and root system

c) Leaves that had dried up because of the uprooting.
d) Small, coloured fruits that originated from flowers. They were wrinkled and dried up on the tree like the leaves; within these fruits there were pips: seeds.
Probable age: 20 years.

1.1.2 On a Scotch pine from the Fontainebleau forest (*Pinus sylvestris*) we observed:

a) Roots that had grown superficially, but there was also a vertical main root.
b) A bare reddish trunk 20 m long with large longitudinal cracks in the bark.
The branches at the top were spreading and the lower branches had fallen off.
c) Pine "needles" joined together in twos in sheaths attached to the branches in place of leaves. It may be noted that the tree had retained its leaves.
d) Cones, directed downwards, of various shapes and colours, depending on their age and sex (we will see

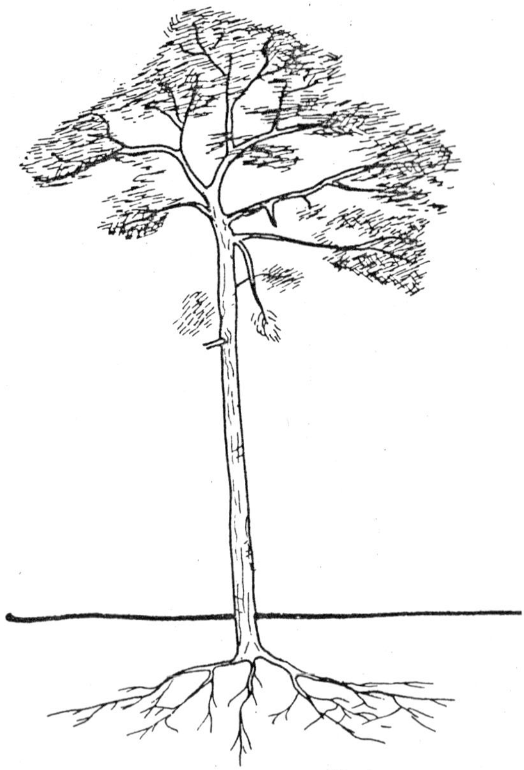

Fig. 2.2. Schema of a Scotch pine

why, in a study of reproduction in conifers) with seeds inside them.
Probable age: 40 years.

Fig. 2.3. A tagetes plant flowering in the summer

1.1.3 The French marigold (*Tagetes patula*) sold by horticulturists for planting in summer beds in our gardens, when uprooted, revealed:
a) A mass of soil retained by many small tangled roots.
b) Several branched herbaceous stems, 10 cm long, bearing:
c) Leaves that were opposite and denticulate, with a characteristic odour.
d) Yellow flowers that looked like small pompoms (further on, we will see that each of these is a collection of flowers on a common receptacle).
Planted in the spring, its vegetative cycle ended before the winter.
Age: less than one year.

1.1.4 We uprooted a wheat plant (*Triticum sativum*) in a field before the harvest.
We observed:
a) A tuft of many roots that held some soil. Above the ground there were:
b) several long, narrow, hollow cylindrical stems with some joints and bearing:
c) on each of them a ribbon-like leaf;
d) at the tip, an "ear of wheat", a spike composed of seeds that originated from flowers, which were themselves hardly visible during flowering.
This spring wheat was sown in March and was harvested during the summer.
It completed its life cycle in one season.

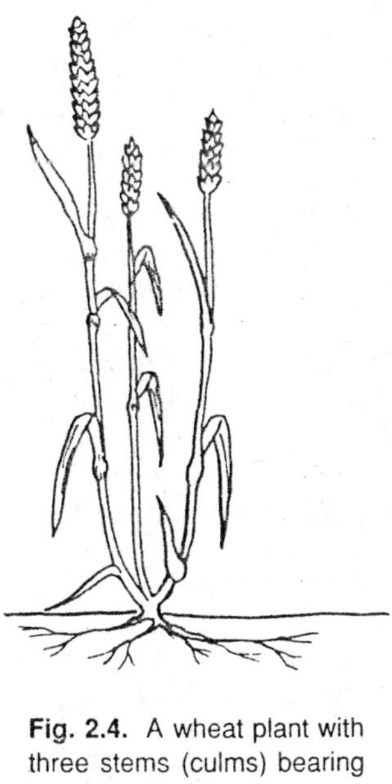

Fig. 2.4. A wheat plant with three stems (culms) bearing leaves and spikes

1.1.5 The myosotis seed (*Myosotis alpestris*) sown in June produced a sapling which, planted in August, was uprooted in the autumn.
It had:
— tangled roots;
— a branched stem;
— single, somewhat elongated leaves but the blue flowers did not appear till the next spring and it may be noted that several flowers were gathered on the same flowering branch.

The myosotis, therefore, has a two-year life cycle. When seeds are formed in its second year, the plant dies.

Fig. 2.5. A myosotis plant flowering in the spring

Fig. 2.6. A tulip plant with its bulb, leaves and single flower in the spring

1.1.6 Present-day propagation in horticulture is not restricted to the use of seeds as in the case of:

— the Scotch pine;
— the French marigold;
— wheat;
— the myosotis.

Tulips (*Tulipa gesneriana*), for example, are grown from bulbs (except when conducting research for new varieties: see Chapter 7, *Genetics* section on "hybridisation").

A bulb planted in November will, in the spring, produce:
a) lanceolate leaves,
b) a single flower supported on a long floral peduncle or scape.

After flowering, the new uprooted bulb has many short fine roots at its lower end.

When the tulip flower is allowed to ripen we can collect seeds from within the fruit.

It takes two years to obtain these seeds. If, however, we replant the bulb collected after flowering in the autumn, a new cycle will begin without seeds.

1.1.7 Finally, a lupin plant (*Lupinus polyphyllus*), above the ground, has:

— a tuft of compound leaves in the shape of a palm with fingers:

— flowers in clusters; seeds in fruits that are shaped like pods (as are those of the pea).

In the soil, a perennial stock perpetuates the plant every year because the aerial parts disappear in the winter.

Fig. 2.7. A tuft of lupin from a perennial underground stock with three flowering branches

1.2 Common characteristics and description of organs

In these six varied horticultural examples we find:

— underground organs that attach the plant to the soil: *the roots;*

— aerial or underground organs: *stems* that are intermediary between the roots and leaves;

— green organs on stems: *the leaves;*

— reproductive organs: *flowers* that usually produce *fruits* and, later on, *seeds.*

Such organs are generally seen on all flowering plants. There are, however, specific cases in which one of these may not exist.

Plants without stems are: acaulous plants;

Plants without fruits: the Scotch pine.

It may be noted that if a plant has flowers, there will normally be seeds.
A seed is the origin of a plant whatever the final stage. From the little begonia that is 6 cm tall to the oak that is 30 m high, the process is always the same.
Let us, therefore, study the morphology of these different organs.

1.2.1 The root

This is the organ that holds the plant in the soil and also draws minerals that it uses for nutrition.
If we pull out a plant we uproot it with some soil.[1]
We have observed primary and secondary roots.

1 If a plant has many roots there is a tangle of roots and soil. We then have an uprooted plant (*Begonia gracilis*). If, however, there are few roots, soil will not remain on them and the roots will be bare, as observed in the extreme case of the carrot (*Daucus carota*).

1.2.1.1 Different parts of roots

Let us germinate seeds of the pea (*Pisum sativum*) on black paper in a moist environment for clearer observation.
After a few days a white axis will emerge from each seed; this is the young root or radicle on which, from bottom to top, we identify:
a) A pointed end protected by a cap, a type of covering visible under a magnifying glass.

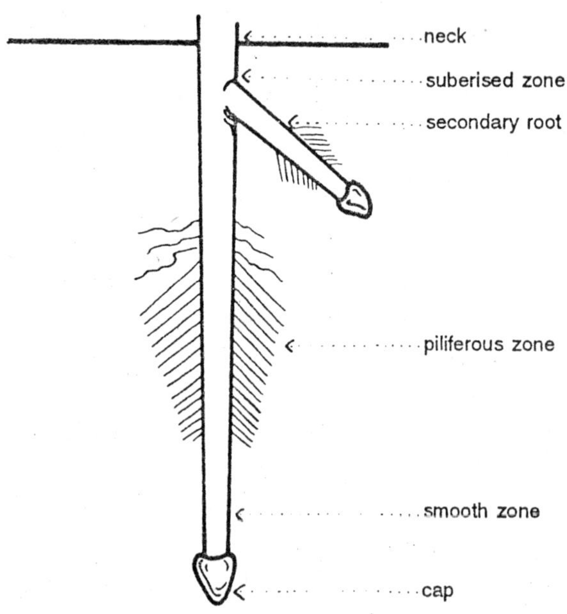

Fig. 2.8. Different parts of a root

b) Above this, a glabrous zone 2 to 3 mm long, separating the cap from the absorbent hairs.
c) Next, a zone covered with these absorbent hairs behind which young lateral roots will appear. This latter zone is called the suberised zone.
Later, the lateral roots grow and become secondary roots.
Each secondary root is, like the primary, composed of
— a cap,
— a smooth region,
— a piliferous zone, and
— a suberised zone.

1.2.1.2 Role of various parts of a root

a) The cap
It covers the tip, known as the "apex" of a root.
It is a type of highly resistant cover, of a dark colour, intended to protect the root against friction as it penetrates the soil.
Its length and thickness vary, depending on the species.
It is formed of layers of cellulose that are renewed as and when they wear out.

b) The smooth region
This is located above the cap and is the centre of increase in length of the root.
Morphological proof:
On a young radicle we draw horizontal lines using Indian ink, 1 mm apart, from the cap to the suberised zone (inclusive). After it grows we observe:
— that the first millimetre, which corresponds, partly, to the cap has hardly increased in length;
— that it is the third and fourth millimetres that have elongated;
— that this increase in length decreases progressively and there is no increase after the tenth millimetre.
This elongation is, therefore, subterminal.
The length of the smooth region remains constant.

c) The piliferous zone
This zone maintains a constant length throughout the life of the root and, further, is always at the same distance from the cap.
Under the microscope we can count several hundred absorbent hairs per square millimetre on a radicle.
Their life is, however, relatively short: only a few days.
In fact, as the root lengthens, the oldest absorbent hairs (upper part) fall off and others emerge at the lower part.

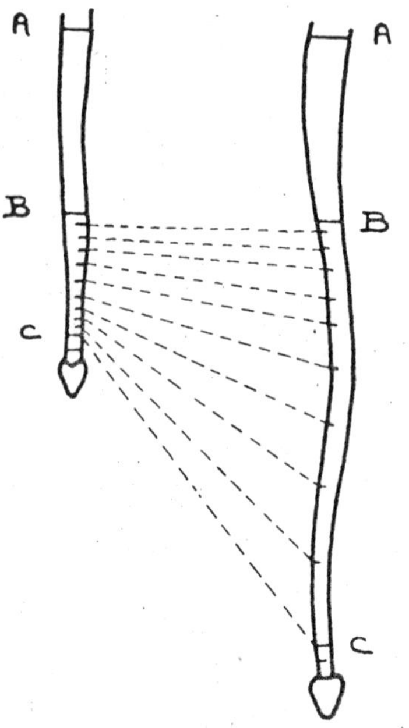

Fig. 2.9. Experiment with "Indian ink lines" indicating subterminal elongation of a root

It may be noted that there are no absorbent hairs on aerial roots.

d) The suberised zone
This is the centre of secondary branches in the root.
e) Finally, the collar, the transition zone between the root and the stem.[2]

2
A root penetrates the soil and its task should be made easier by a ploughing and breaking up that will soften the soil in a field.

1.2.1.3 Different types of roots and their adaptation to the environment

The examples that we selected to observe the morphology of flowering plants have revealed that there can be large main roots with a few secondary ones (as on the pine) or just the opposite (as on wheat).
We are all aware of the roots of ivy (*Hedera helix*) that attach the plant to walls and tree trunks; they are aerial roots.
The aforementioned are only three cases. There are several others.
Let us examine them.

■ *Underground roots*
On the oak (*Quercus sessiliflora*), the salsify (*Tragopogon porrifolius*) and the broad bean (*Vicia faba*), the

secondary roots are much thinner than the primary one.
We, therefore, have tap roots.[3]

3
Tap roots exhaust the soil to some depth.
Stony soils cause tap roots to become forked or twisted.
Before plants with tap roots are grown, the digging has to be deep.

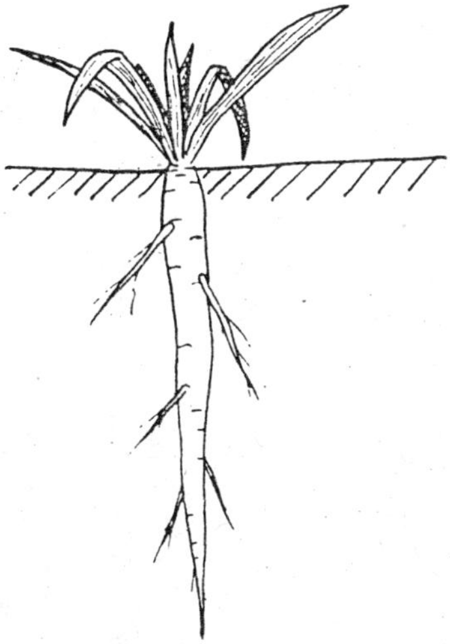

Fig. 2.10. The tap root of the salsify

They penetrate the soil vertically in a better way if there are no obstacles (stones etc.).
In the case of the grass family—wheat (*Triticum sativum*), the castor plant (*Ricinus communis*) and the begonia (*Begonia gracilis*)—the secondary roots are larger than the tap root: they are, therefore, fasciculate roots.[4]
If fasciculate roots remain on the surface and become

4
Fasciculate roots exhaust the surface of the soil.
A fasciculate root system helps in "revival" during planting. This is why, at each planting-out or transplantation in a nursery, the horticulturist cuts the tap root in order to ensure a development of fasciculate roots. This is one reason for trimming.

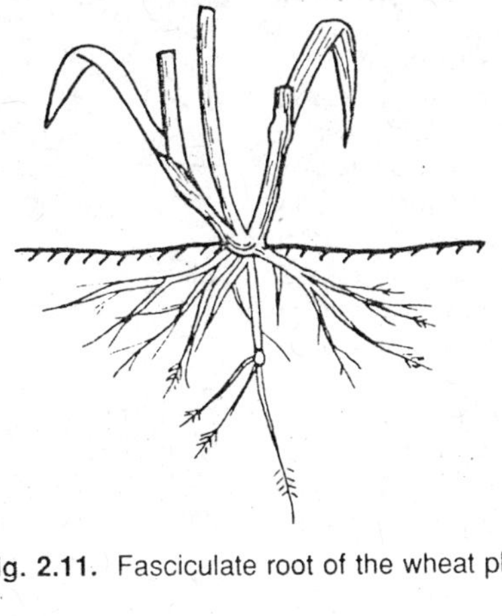

Fig. 2.11. Fasciculate root of the wheat plant

very long they are *creeping roots,* as seen on the vine (*Vitis vinifera*).
A plant often stores reserve material in its roots, which become swollen. They, then, become tuberous roots.[5]

5
Market gardeners grow many tuberous roots as food: red beetroot (*Beta vulgaris*), turnip (*Brassica napus*).

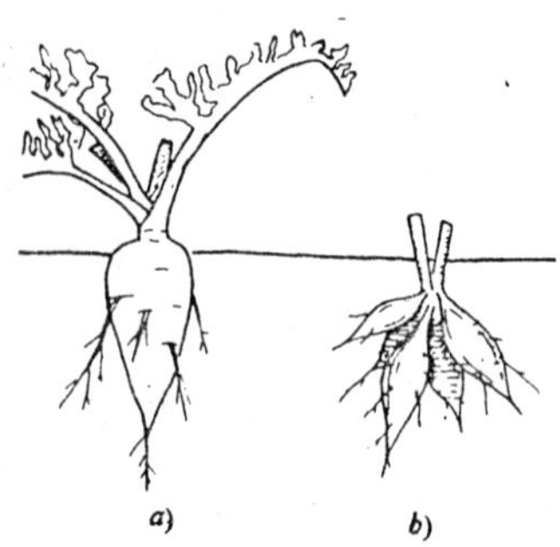

Fig. 2.12. Tuberous roots: (a) tap root of the carrot, (b) fasciculate root of the dahlia

We may observe:
— tuberous fasciculate roots: dahlia (*Dahlia hortensis*).
— tuberous tap roots: radish (*Raphanus sativus*), carrot (*Daucus carota*).

■ *Adventitious roots*
These appear in no particular order, for example:
☐ On creeping aerial stems: the aerial stem of the strawberry (*Fragaria vesca*) has runners called "stolons" that increase in length and on which, at intervals, there are nodes from where leaves emerge. These nodes eventually develop roots.[6]

6
Market gardeners often use this natural layering as a method of propagation for the strawberry plant.

Fig. 2.13. Adventitious roots on a creeping aerial stem of the strawberry

☐ On climbing aerial stems: ivy (*Hedera helix*), for example, has a type of crampons that penetrate cracks in the bark on tree trunks and between stones in walls. These crampons eventually develop as roots in a favourable environment.[7]

7
While, in ivy (*Hedera helix*), crampon-roots enable the plant to climb, this damages cement coatings.
For this reason, and in certain cases, we prefer the Virginia creeper (*Ampelopsis veitchii*), whose "adhesive dises" are modified leaves and remain on the surface. Certain plants of tropical or equatorial origin have aerial roots that hang: philodendron (*Philodendron pertusum* or *Monstera deliciosa*), orchids (*Cattleya, Vanda, Cymbidium*).
Since their role is to fix atmospheric humidity they should not be cut or exposed to a very dry atmosphere.

Fig. 2.14. Adventitious roots on a climbing aerial stem of ivy

8

This possibility is often availed of by the horticulturist.

- Graminaceous plants in a lawn that is rolled to facilitate "trimming".
- Beans (*Phaseolus vulgaris*) that are ridged, thereby facilitating the development of a root system.
- The India rubber tree (*Ficus elastica*), which is layered.
- A large number of plants that are grown from cuttings.

In some cases, leaves (*Begonia rex, Saintpaulia ionantha*) act as stems. It may be noted that the development of adventitious roots is facilitated by the use of various chemical products (see Chapter 6, section on *asexual reproduction*).

☐ On underground stems (or stems planted in the soil): a buried stem may, in certain circumstances, develop adventitious roots: couch-grass (*Agropyron repens*), iris (among others: *Iris kaempferi, Iris pumila*).[8]

Fig. 2.15. Hanging aerial roots of the philodendron

■ *Aquatic roots*

Certain plants have their roots in water.

In these the roots are fairly long and, in any case, do not have absorbent hairs: duckweed (*Lemna gibba*),

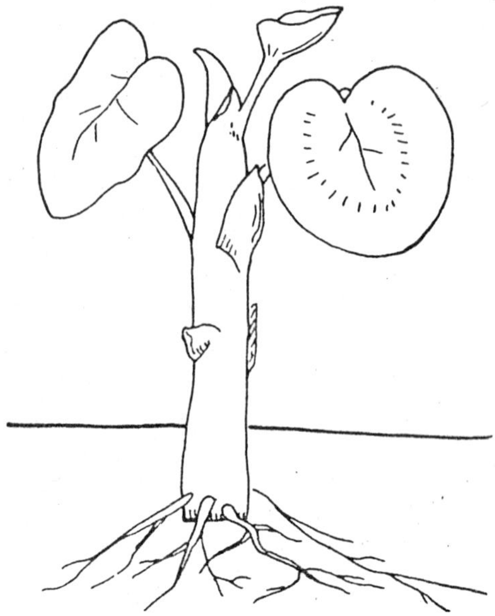

Fig. 2.16. Adventitious roots on an aerial stem planted in the ground; a cutting of pelargonium

water lily (*Nymphaea*), watercress (*Nasturtium officinale*).
In watercress, adventitious roots develop from axillary buds on aquatic stems.

■ *Epiphytic roots*
In some tropical orchids the roots attach the plant to branches of trees that act as supports. This is not parasitism (*Platycerium alcicorne*).

■ *Endophytic roots*
These are roots of parasitic plants that attach them to hosts to draw the sap they need.
—for example, mistletoe (*Viscum album*) on the apple tree, dodder (*Cuscuta*) on lucerne.

1.2.2 The stem

The stem is the aerial part of a plant bearing leaves and reproductive organs (flowers, fruits, seeds).
It is an organ with chlorophyll (green), at least in the case of annual stems.

1.2.2.1 Different parts of a stem

a) The demarcation between the root and the stem is called the *collar.*[9]

9 By definition the collar will be at ground level.

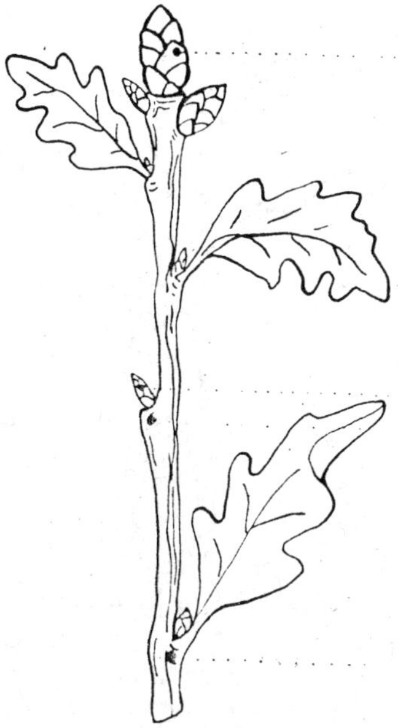

Fig. 2.17. Different parts of a stem

b) The space between two leaves is an internode and, by definition, an internode never has leaves. Internodes become progressively shorter as we near the top of the stem.

c) When the internodes are reduced to a minimum, the leaves are inserted one beside the another. They cover each other: this is a bud.

d) At the end of the stem we have a terminal bud.[10]

e) In the axil of each leaf we have:
—*an axillary bud* that, sooner or later, gives rise to branches with leaves; wood or flower buds: fruit buds.[11]

f) Since leaves are arranged in a specific order on stems (*see chapter on phyllotaxy*), axillary buds, are arranged in the same order.
Others, called *adventitious buds,*[12] are not arranged in any particular order.

g) A bud usually produces a young stem only in the year after its formation. As a result of an accident (broken branch), the highest bud before the break (sometimes the two or three upper buds) may develop

10

Usually small, it may be enlarged and filled with food reserves: cabbage (*Brassica oleracea capitata* or *bullata*).

11

a. The arboriculturist would like to ensure a development of axillary buds:
— as early as possible,
— as flower buds and not vegetative buds.
This is one of the aims in pruning fruit trees.
b. The market gardener benefits from the enlargement of axillary buds on Brussels sprouts: numerous small leaves.

12

a. The horticulturist may, by pruning, cause adventitious buds to develop. He alters the shape of the plant: ornamental pruning.
b. A strong stem when cut down will produce adventitious buds under the cut point: pruning of the plum tree (*Prunus domestica*), trimming of the catalpa (*Catalpa bignonioides*).
c. Adventitious buds may appear on roots (suckers); on leaves (leaf cuttings); on branches (epicormic stems).

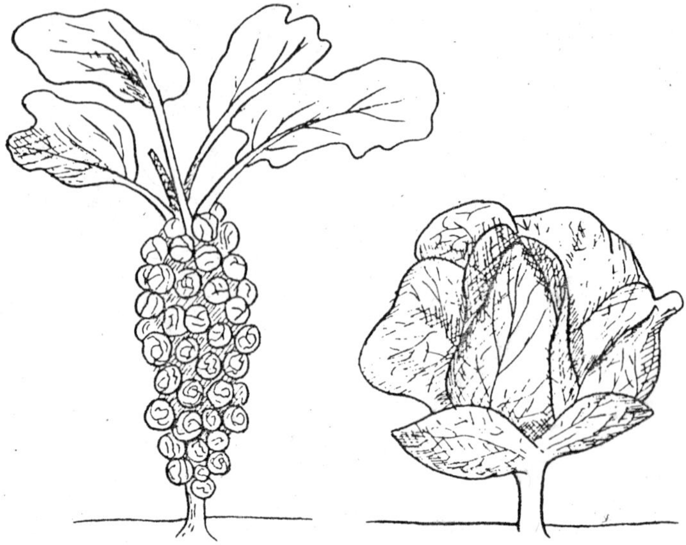

Fig. 2.18. Enlarged axillary buds of Brussels sprouts (left). The enlarged terminal bud of the cabbage (right).

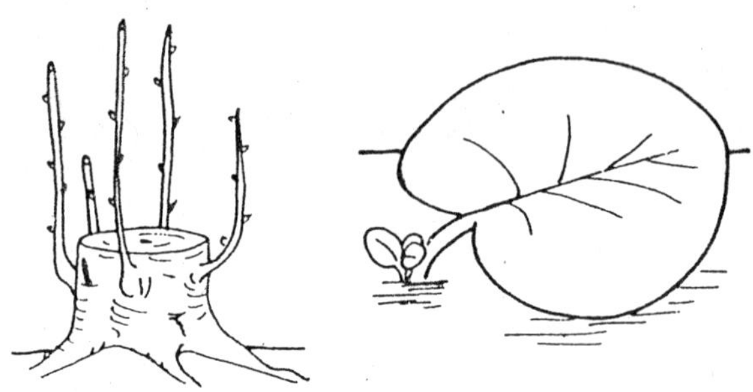

Fig. 2.19. A strong stem when cut down enables the development of adventitious buds: pruning of the plum tree (left). Emergence of adventitious buds on a leaf cutting of saintpaulia (right).

during the same year. These are known as anticipated buds.[13]

13

By pruning and pinching during growth, the arboriculturist works on the development of these adventitious buds (*see the course on growing fruit trees*).

1.2.2.2 Role of various parts of a stem and its elongation

The stem serves as a liaison between the roots and the leaves (see Chapter 5, *Plant Physiology*, §1.5, Circulation of sap).

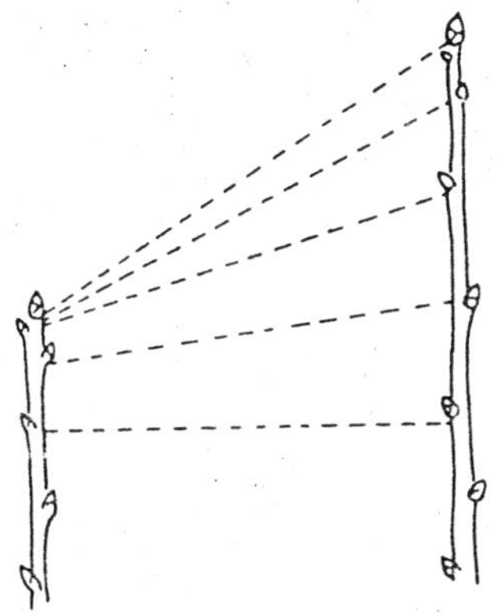

Fig. 2.20. Terminal and intercalary elongation of a stem

Buds are branches in an embryonic state. If they are to remain as such in the winter (on trees, shrubs, perennial plants, etc.) they are covered with protective scales. We can observe easily that the elongation of a stem is different from that of a root.

a) The terminal bud opens out, the young leaves that form it extend, and the point (apex) of the stem is lengthened.

There is, therefore, a *terminal elongation.*

b) The internodes that are, initially, short in the bud gradually increase in size till they attain their final lengths.

There is also an *intercalary elongation.*

A plant without a stem is said to be acaulous—leek (*Allium porrum*), dandelion (*Taraxacum dens-leonis*), carrot (*Daucus carota*)—at least till flowering, when a stem grows to bear flowers.[14]

14

The most recent horticultural applications are based on controlling stem elongation using growth regulators the effects of which develop in two ways.

a. By inhibiting the elongation of internodes with "dwarfing" products such as

— chlormequat (e.g. Cycocel).
— daminozide (e.g. Alar).
— paclobutrazol (e.g. Bonzi).

Plants in beds and some potted plants are treated in this manner (hortensia, *Hydrangea macrophylla;* impatiens of New Guinea, *Impatiens hawkeri;* gardenia, *Gardenia jasminoides;* hibiscus, *Hibiscus rosa-sinensis*). It is clear that a "dwarfer" has to be used on young organs as soon as the shoots begin to grow (e.g. shoots of 2 to 3 cm).

b. By reduction of the apical dominance that facilitates development of axillary shoots. The plant is shorter and becomes more tufted. In the extreme case of destruction of an apical meristem we will have carried out an actual "chemical pinching".

One of these regulators is dikegulac (e.g. atrinal).

Its use is limited in France and is observed on fuschia (*Fuchsia ricartonii*). It is more frequent in Belgium on azalea (*Azalea indica*).

Growth regulators are applied by spraying or watering.

They are absorbed through the leaves or roots.

The cost of interventions is relatively limited but the use of these treatments is not yet widespread.

An excess of the active product is likely to cause a stoppage of growth. An insufficient amount is ineffective. While treatments in

14 contd.

greenhouses can still be controlled this is not possible in open-air beds. A draining of the product due to rain may result in considerable persistence in the soil which creates a problem for its application on hedges.

Another example: growth regulators when applied on lawns render it difficult to obtain good results because they affect a number of species of grasses that are not always at the same stage of growth. The results thus vary greatly from one species to another and end in a disordered state of the lawn.

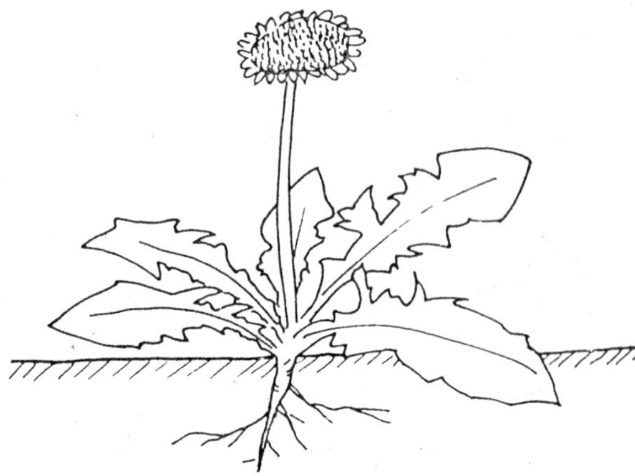

Fig. 2.21. Dandelion acaulous plant

Fig. 2.22. Trial of two new growth reducers (Bonzi and Exp. 30054) on garden plants (*Pelargonium zonale* "top score") (CNIH document)

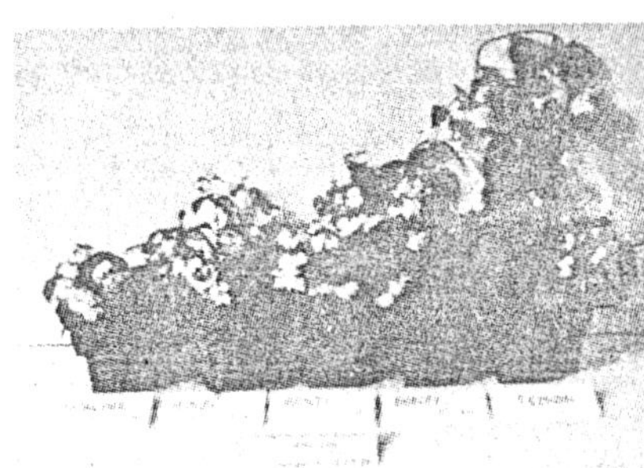

Fig. 2.23. Trial of a new growth reducer (Bonzi) on *Clerodendron thomsonae* (CNIH document)

1.2.2.3 Different types of stems and their adaptation to the environment

As in the case of roots, stems are adapted to the environment and may be classified in three categories:

— those above the soil,

— those in the soil,
— those in water.

a) Aerial stems

These stems are usually erect.

In this group we have:

— herbaceous stems,
— woody trunks of trees,
— unbranched candices of palms,
— hollow-cylindrical culms of cereals.[15]
— Some aerial stems are trailing, like the stolons of the strawberry (*Fragaria vesca*).
— Others may be climbing stems.

In such cases there are several modes of fixation of the stem:

— either by tendrils: the pea (*Pisum sativum*), the vine (*Vitis vinifera*),
— or by crampon roots: ivy (*Hedera helix*),
— or by adhesive discs: Virginia creeper (*Ampelopsis veitchii*),
— or by twining around a support: bindweed (*Convolvulus*), wisteria (*Wisteria*).

It may be noted that these so-called twining plants always twine in the same manner.

The woodbine (*Lonicera caprifolium*) The polygonum (*Polygonum baldschuanicum*) The Japanese wisteria (*Wisteria floribunda*)	From right to left (Senestrorsum)
The jasmine (*Jasminum fruticans*) The Chinese wisteria (*Wisteria sinensis*)[16]	From left to right (Dextrorsum)

b) Underground stems

We observe three distinct groups of such stems:

☐ *Rhizomes:* These are horizontally elongated stems with lateral roots. At the end of each rhizome a bud produces the annual aerial part; when it falls, a scar remains. This is observed in the Iris (*Iris kaempferi*), the lily of the valley (*Convalaria majalis*), the fern (*Polypodium glaucum*), and couch-grass (*Agropyron repens*).[17]

☐ *Tubers*: These are swollen stems filled with reserve materials. Proof that they are underground stems and not roots lies in the existence of eyes on the tuber, arranged in a regular manner:

Potato (*Solanum tuberosum*),
Jerusalem artichoke (*Helianthus tuberosus*),
Japanese artichoke (*Stachys sieboldii*).[18]

15

It is the height and starting point of the branches on trunks that enable us to classify certain plants.

— If a trunk is more than 5 m high and the first branches begin at a height of more than 2 m the plant is a *tree:* oak (*Quercus*), beech (*Fagus*).
— If the branches begin from the ground and the plant is 2 to 5 m high it is a *shrub*: cherry laurel (*Prunus laurocerasus*).
— If the branches begin from the ground and the plant is not more than 2 m high it is a *bush* (*Spiraea* × *vanhouttei*).

16

The method of fixation of climbing stems is based on the necessity, or otherwise, of a support.

We do not support the Virginia creeper (*Ampelopsis orientalis*) or ivy (*Hedera helix*).

We support the pea (*Pisum sativum*) with props.

We support the wisteria (*Wisteria sinensis*).

17

Since rhizomes may develop adventitious buds, in certain cases, this may result in new individual plants, a natural development of a stem cutting. This is observed in couch-grass, of which all fragments of rhizomes must be extracted to ensure an effective clearing of the soil.

18

These tubers filled with food reserves are cultivated by floriculturists, market gardeners and agriculturists.

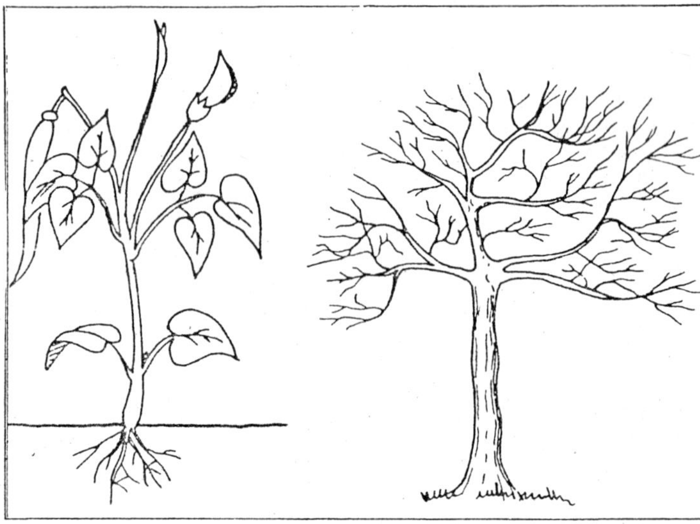

the herbaceous stem of the haricot bean — the trunk of the oak

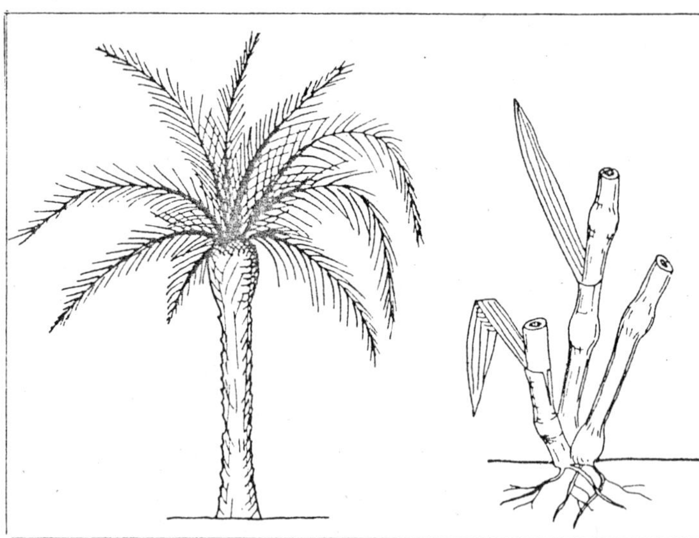

the palm caudex — the culm of wheat

Fig. 2.24. Erect aerial stems

Fig. 2.25. The climbing stem of the Virginia creeper, held by adhesive discs

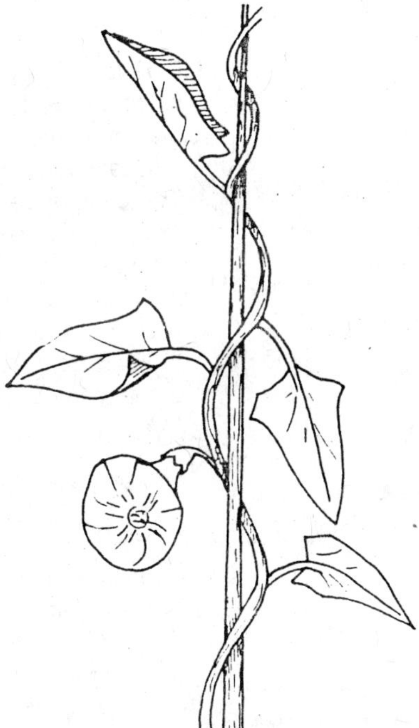

Fig. 2.26. The climbing, twining stem of the convolvulus

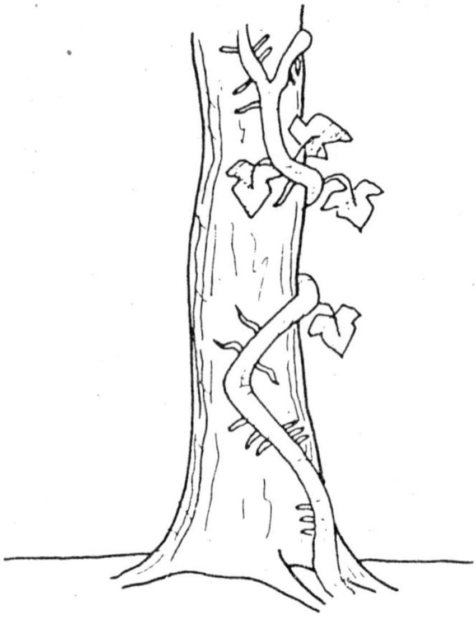

Fig. 2.27. The climbing stem, attached by crampon roots, of ivy

□ *Bulbs*: These are short stems covered with imbricated leaves. Depending upon the organ (stem or leaf) that stores reserve materials, we observe:

1) *Scaly bulbs*: reserves in swollen stems.

The base is perennial and, on its lower surface, has adventitious roots. The leaves are transformed into thin scales that are formed by the annual development of the bud. They are also separate from each other as seen in the lily (*Lilium candidum*).

2) *Tunicate or foliate bulbs*: in which the development of the stem is limited.

— A vegetative unit lasts for a year.

— The base has a terminal bud, many adventitious roots and thick leaves that act as reserves for the year.

A new vegetative unit is formed from a bud in the axil of a scale and the older one is resorbed: for example, the onion (*Allium cepa*) and the tulip (*Tulipa X*).[19]

Sometimes, axillary buds develop to form cloves: for example, garlic (*Allium sativum*).

3) *Solid bulbs, or corms,* that appear as swollen stems, or a type of very short rhizomes, resembling bulbs: for example, the gladiolus (*Gladiolus*) and saffron (*Crocus sativus*).

19 In the case of tunicate bulbs of the tulip, the bulb uprooted after flowering is not the same one that had been planted in the autumn.

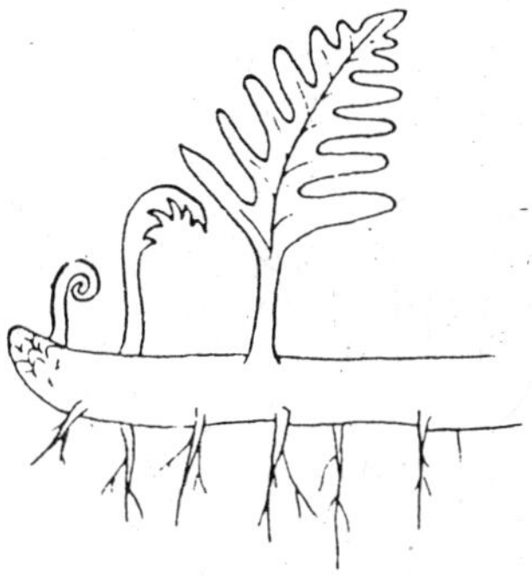

rhizome of the polypody

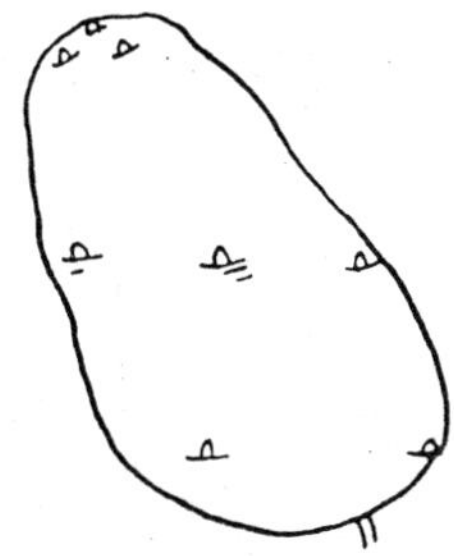

tuber of the potato

scaly bulb of the lily

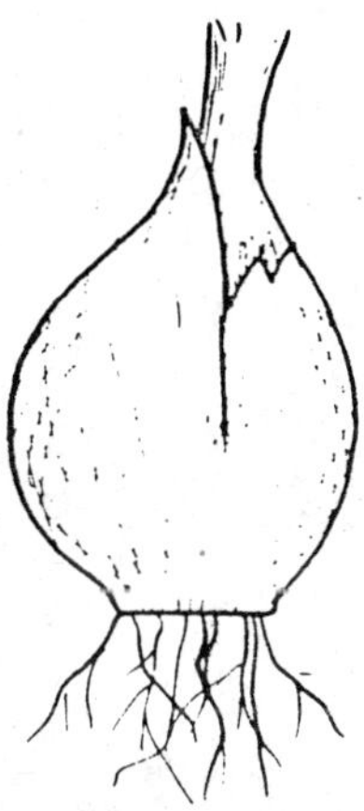

tunicate bulb of the tulip

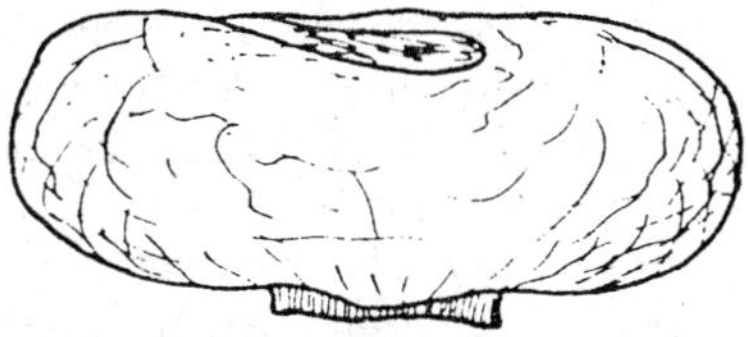

solid bulb or corm of the gladiolus

Fig. 2.28. Underground stems

c) Aquatic stems
These stems have cavities containing air that enable them to float: for example, the water lily (*Nymphaea alba*, among others).
Note: Some stems may have an altered structure to cater for reserves of water, essential for the plant during certain dry periods: for example, cacti.

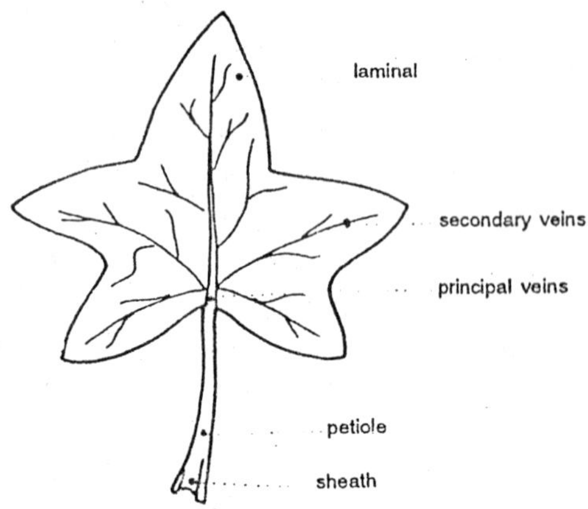

Fig. 2.29. A leaf and its parts

1.2.3 The leaf

1.2.3.1 Different parts of a leaf

a) The *lamina* is formed mainly by a flat green blade, somewhat serrated and supported by venations.
It is generally of a *darker* colour on the upper side, which is exposed to the sun, than on the lower side (myosotis).
There are veins, all of the same size (wheat), or there may be primary and secondary veins: myosotis, the French marigold (*Tagetes*).
In some cases the leaves are reduced to the state of needles: Scotch pine (*Pinus sylvestris*).
A lamina may be in a single piece or have several parts (*see "Shapes of leaves"*). The leaf is then compound and may have several leaflets (lupin).
We identify a leaflet by the fact that it never has axillary buds at its base.
b) A lamina is supported by a "tail", *the petiole*. This is not always present: *the leaf is then sessile*: woodbine (*Lonicera sempervirens*), iris . . .
On a compound leaf, the leaflets may also be sessile.
c) The petiole may be dilated at the base to form a sheath that encircles the node on the stem.

If the sheath completely surrounds the stem, *the leaf is a sheathing one*: for example, wheat (*Triticum sativum*).

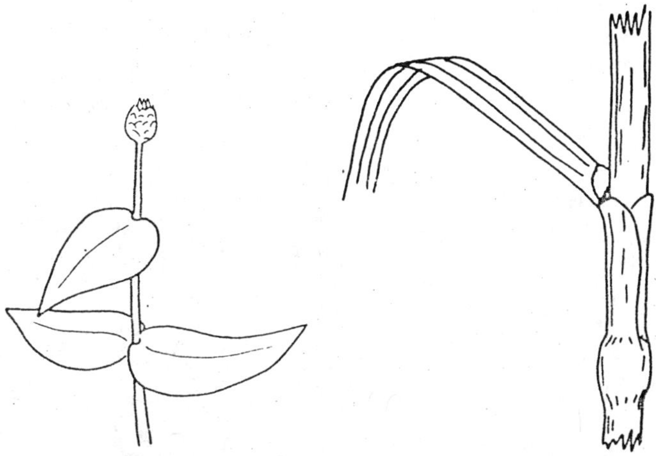

Fig. 2.30. The sessile leaf of the zinnia (left). The sheathing leaf of wheat (right).

d) The sheath may, sometimes have two green blades, the stipules: dog rose (*Rosa canina*), strawberry (*Fragaria vesca*). These stipules may persist with the leaf or disappear after it opens out.

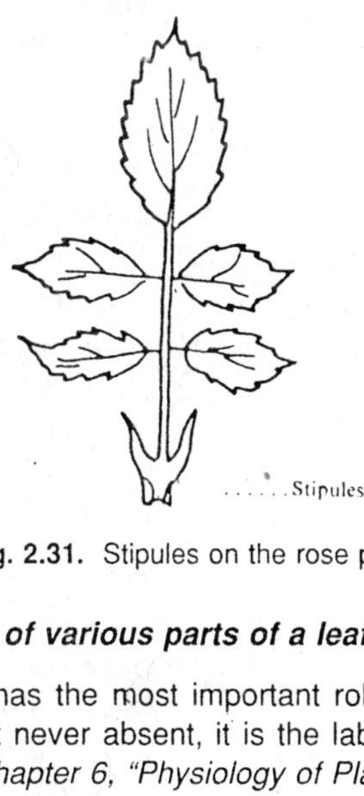

Fig. 2.31. Stipules on the rose plant

1.2.3.2 Role of various parts of a leaf

The lamina has the most important role. Sometimes modified, but never absent, it is the laboratory of the plant (*see Chapter 6, "Physiology of Plants"*).

The lamina is always oriented to receive the maximum amount of light.

Leaves have a terminal and intercalary growth.[20]

20

The horticulturist, most often, tries to ensure the maximum leaf surface necessary for the plant.

1.2.3.3 Various types of leaves

The leaves are parts of a plant that vary the most and there are several criteria that enable their classification.

Those are:

a) Their shapes. They may be:

—*Simple*: American sycamore (*Platanus occidentalis*), willow (*Salix babylonica*).

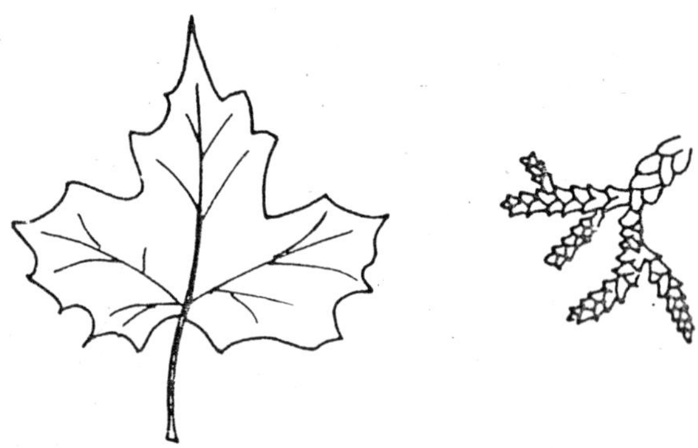

Fig. 2.32. A simple leaf of the sycamore (left). Scaly leaves of the thuja (right).

— *Scaly* (*Thuja lobbii*).

— *Compound*: with several leaflets.

These compound leaves are:

— *Palmate*: if the leaflets are inserted at the same point: horse chestnut (*Aesculus hippocastanum*).

— *Pinnate*: if the leaves are opposite, in pairs.

Compound pinnate leaves may have:

— leaflets in even numbers: leaves that are *paripinnate* (*Gleditsia triacanthos*);

— leaflets in odd numbers (an additional leaflet known as the apical leaflet, at the tip), as in *imparipinnate* leaves: walnut (*Juglans regia*), dog rose (*Rosa canina*).

b) Their positions

Leaves are arranged on the stem in a specific manner: this distribution is known as *phyllotaxy*. It depends on the action of young tissues (meristems) in the buds.

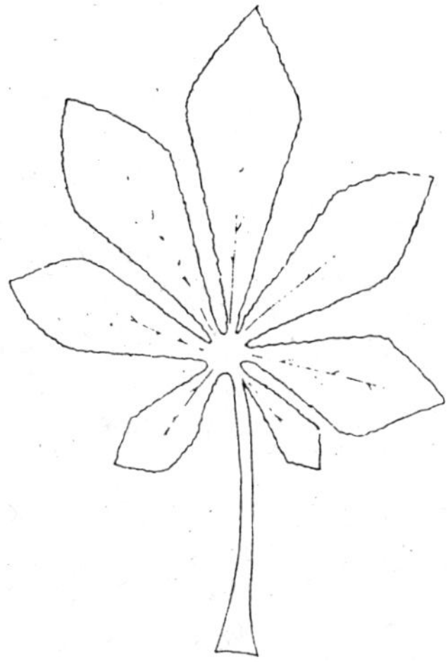

Fig. 2.33. The palmate leaf of the horse chestnut

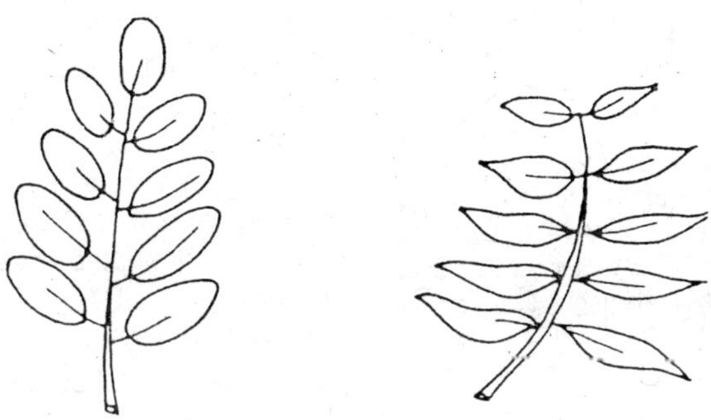

Fig. 2.34. (left) An imparipinnate leaf of the acacia. (right) A paripinnate leaf of the gleditsia.

Separate leaves are inserted singly on a stem: they are alternate: tilia (*Tilia argentea*).

Opposite leaves are arranged facing each other. They are at 90° to each other: leaves in the form of a cross (2 × 2): sage (*Salvia splendens*). Leaves may be inserted in groups of more than two at the same height:

- If they are arranged all around the axis they are verticillate: rose laurel (*Nerium oleander*).
- If they are joined at the same point they are fasciculate: barberry (*Berberis stenophylla*).

Most often they are alternate.

If we join the base of the petioles with a thread leading to the leaves next above them, we obtain a spiral. To

find two leaves that are located vertically, one above another, involves:
x turns of the stem (1 turn = 360 degrees) and y leaves. The relationship x/y gives the angle between two successive leaves. This is the divergence of leaves.
We thus observe the following relationships:
1/2: wheat (*Triticum sativum*), elm (*Ulmus americana*), tilia (*Tilia argentea*).
1/3: birch (*Betula alba*).
2/5: most frequent case: oak (*Quercus sessiliflora*), apple (*Malus communis*), pear (*Pyrus communis*), plum (*Prunus domestica*).
There are also other relationships, such as 3/8, 5/13. In this way leaves receive the maximum amount of light.
c) Their shape
There is considerable diversity in the shapes of leaves. We will deal with only the most common types:
Accuminate (tapering to a point): birch (*Betula alba*).
Elongate: viburnum (*Viburnum rhithydophyllum*).
Cordate (heart-shaped): tilia (*Tilia argentea*), violet (*Viola cornuta*).
Digitate (finger-shaped): lupin (*Lupinus polyphyllus*).
Elliptical: lily of the valley (*Convallaria majalis*).
Ensiform: (sword-shaped): iris.
Lanceolate: rose laurel (*Nerium oleander*).
Linear: carnation (*Dianthus caryophyllus*).
Oblong (elongated oval): privet (*Ligustrum ovalifolium*).
Ovate: pear (*Pyrus communis*), etc.

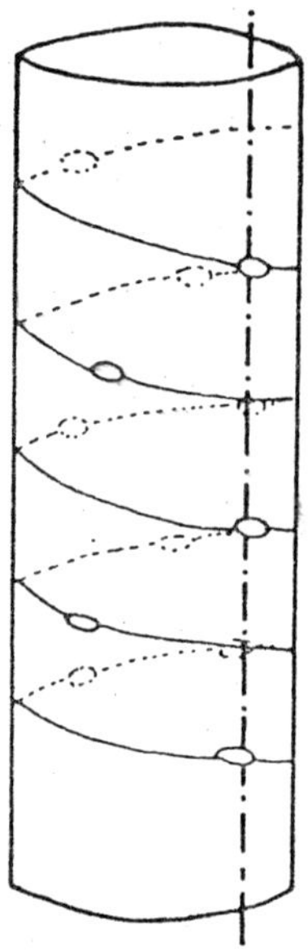

Fig. 2.35. Phyllotaxy
The apple. To determine a leaf that lies vertically above the initial leaf, count 5 leaves while making 2 turns on the stem, that is 2/5.

It may be noted that the same plant may, sometimes, have leaves of clearly different shapes: this is what is called polymorphism in leaves: sycamore (*Platanus occidentalis*).
d) Their margin
As varied as leaf shapes, margins are shallow or deep.
Entire (without serrations): the India rubber tree (*Ficus elastica*), lilac (*Syringa vulgaris*), beech (*Fagus sylvatica*).
Dentate (small notches): stinging nettle (*Urtica dioica*), chestnut (*Castanea sativa*), poplar (*Populus nigra*).
Sinuate (small rounded lobes): oak (*Quercus sessiliflora*).
Lobed (deep lobe in half the limb): ivy (*Hedera helix*), sycamore (*Platanus occidentalis*).
Crenate: violet (*Viola cornuta*).
Incised (irregular incisions): chrysanthemum (*Chrysanthemum indicum*), etc.

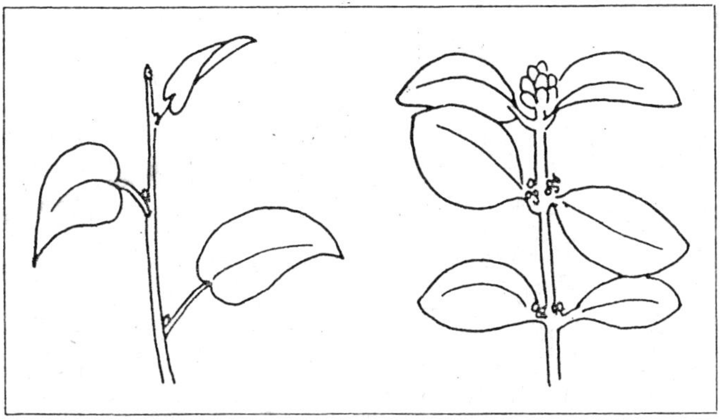

alternate leaves of the pear tree

opposite leaves of the box tree

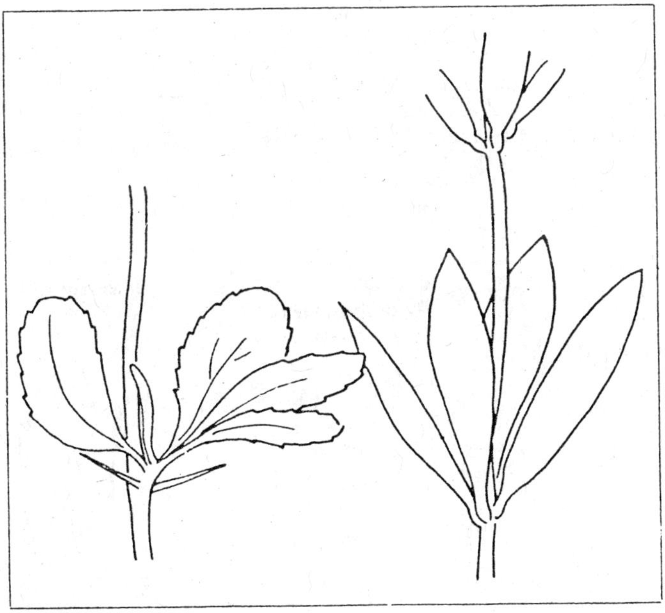

fasciculate leaves of the barberry

verticillate leaves of the rose laurel

Fig. 2.36. Positions of leaves

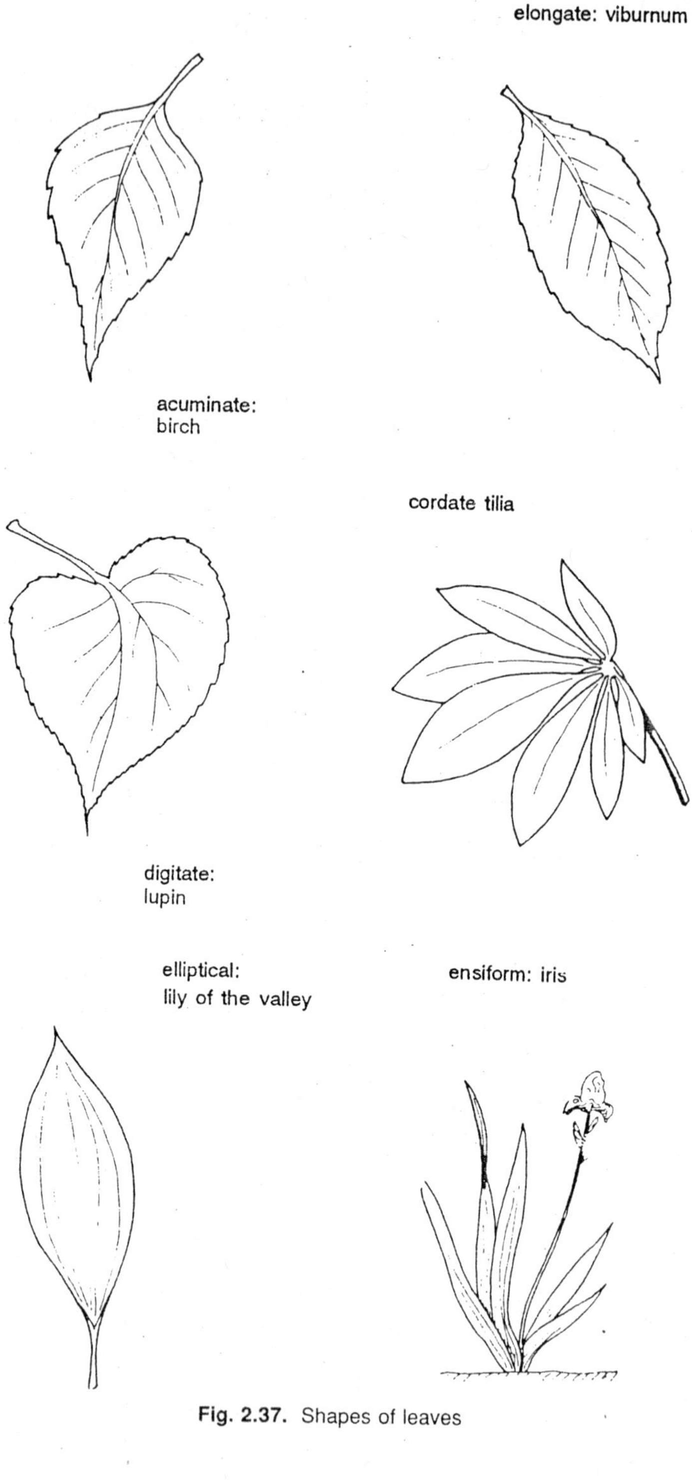

Fig. 2.37. Shapes of leaves

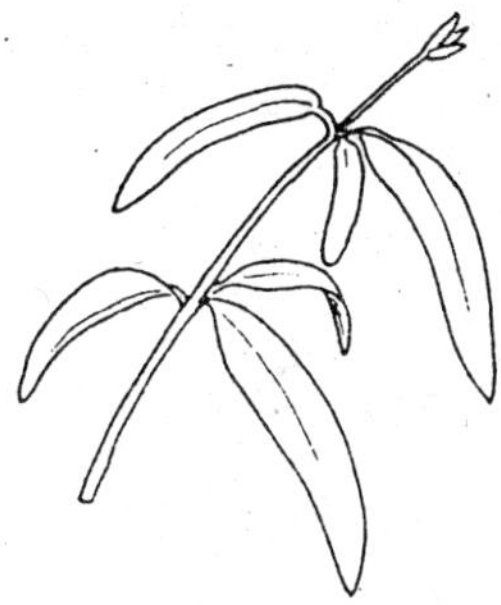

lanceolate:
rose laurel

linear:
carnation

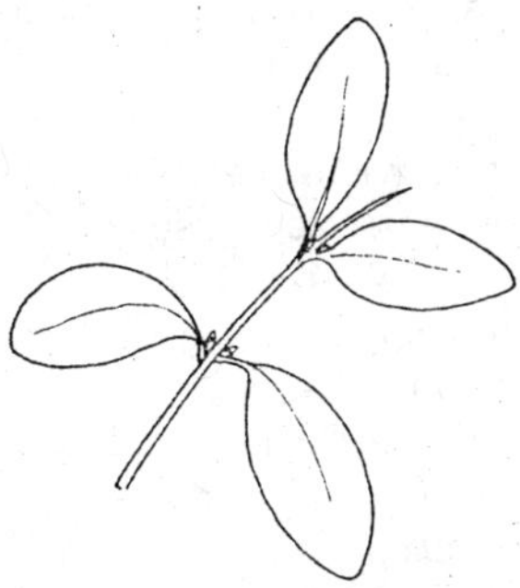

oblong:
privet

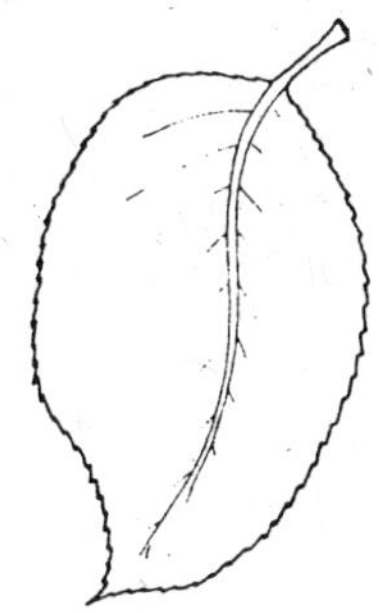

ovate:
pear

Fig. 2.38. Shapes of leaves

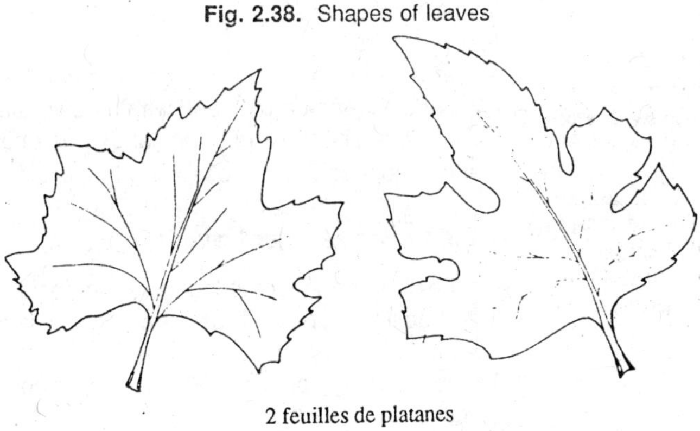

2 feuilles de platanes

Fig. 2.39. Polymorphism in leaves: Two leaves of sycamore

e) Their venation

This refers to the arrangement of veins on the lamina. It may be pinnate (fishbone): pinninervate leaf, the most frequent case: cherry (*Prunus cerasus*), oak (*Quercus sessiliflora*), pear (*Pyrus communis*).

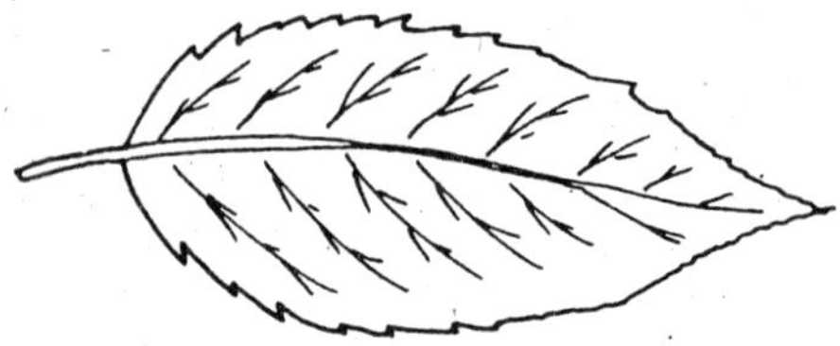

Fig. 2.40. The pinninervate leaf of the cherry tree

It may be palmate: palmatinervate leaf: ivy (*Hedera helix*), horse chestnut (*Aesculus hippocastanum*), maple (*Acer platanoides*), vine (*Vitis vinifera*).
It may be parallel, parallelinervate leaf: monocotyledons (graminaceous plants, etc.).
f) Their insertion on the stem
When there is no petiole we can, especially, identify:
Sessile leaves, without a petiole: cytisus (*Cytisus sessilifolius*).
Sheathing leaves, encircling others: gladiolus (*Gladiolus*), leek (*Allium porrum*).
Amplexicaul leaves, where the lamina encircling the stem is attached to it only at the point of insertion: poppy (*Papaver nudicaule*).
g) Their colour
This very useful characteristic in arboriculture may be obtained in very light shades of green to purple or golden leaves, etc.[21]
h) Their hairs
A leaf without hairs is said to be *glabrous*; if it has a few hairs it is *tomentose*, and with a down, *pubescent*.
Conclusion. These eight principal criteria enable us to identify most commonly observed leaves.[22]

1.2.3.4 Adaptation of leaves

On the other hand, some leaves are modified and lose their traditional appearance, to which we have just referred.
These are actual adaptations due to particular circumstances.
a) Adaptation to the environment
■ *In an aquatic environment*: arrowhead (*Sagittaria sagittifolia*).
The leaves store a reserve of water.
The submerged leaves are shaped like straps.

21
It may be noted that for the so-called ornamental colours we often have species, varieties and, especially, cultivars.

22
Such a variety in the appearance of leaves provides the landscape gardener with a large selection in respect of the effects that he wishes to obtain.

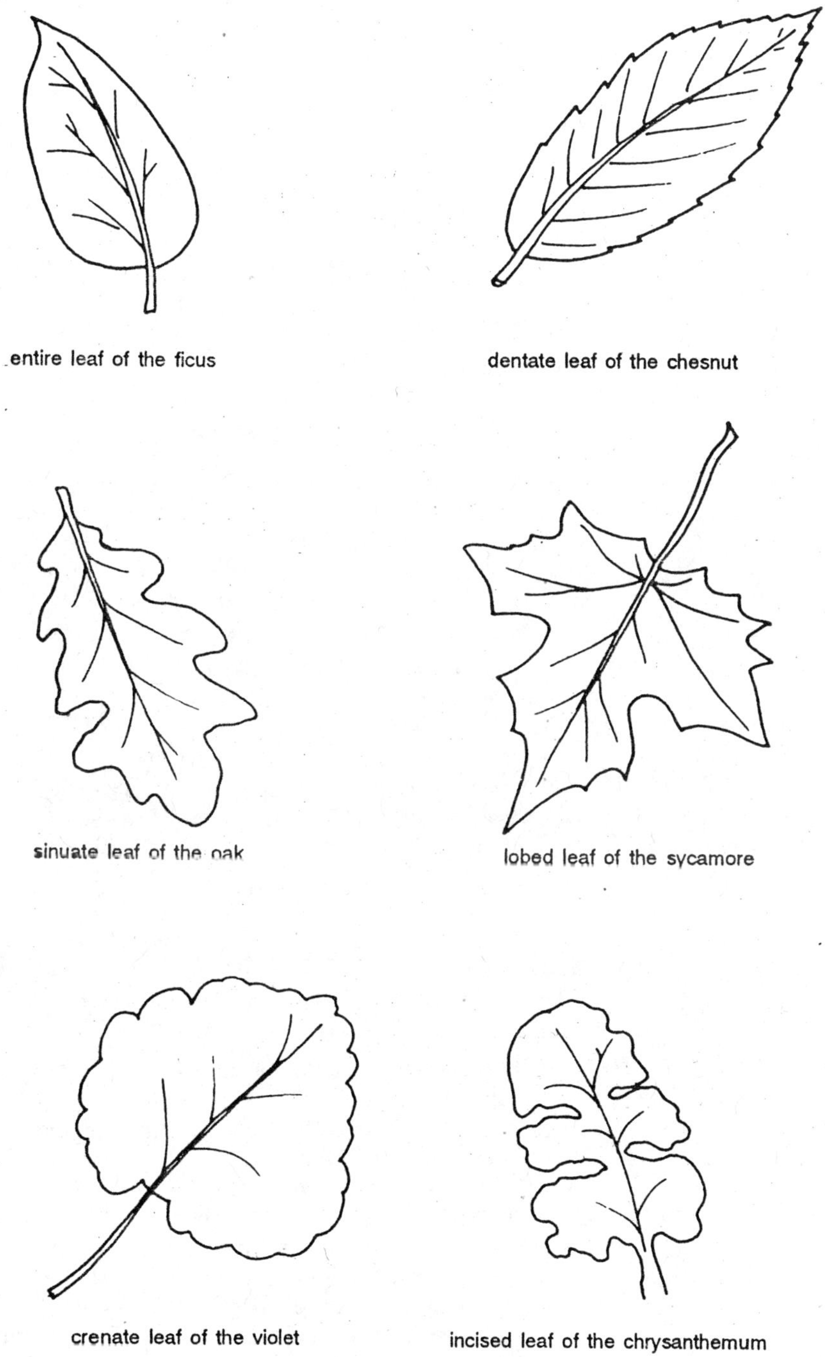

Fig. 2.41. Margins in leaves

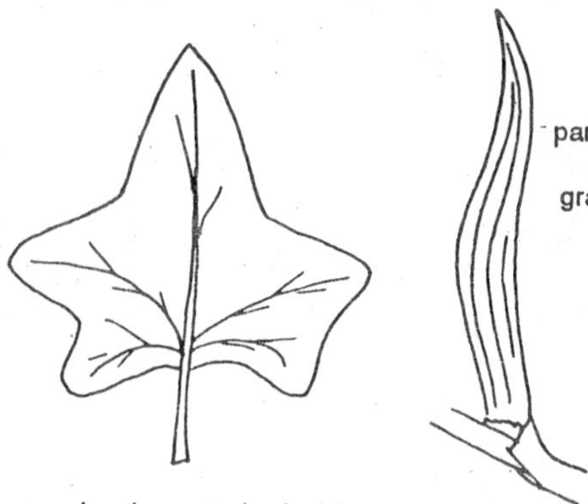

Fig. 2.42. Venation in leaves

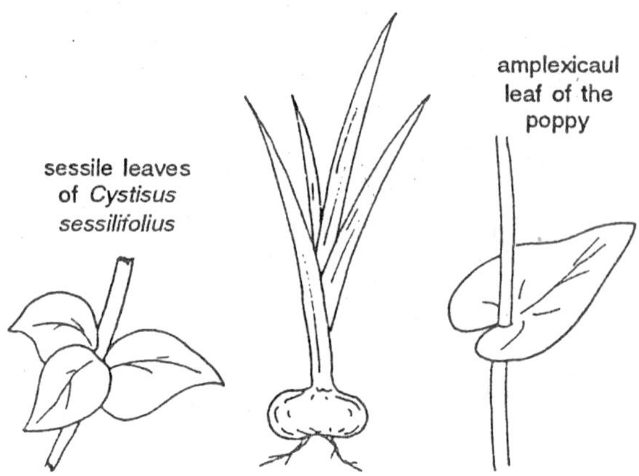

Fig. 2.43. Insertion of leaves on the stem

The leaves above the water surface, being normal, are shaped like spears.

The floating leaves, of a roundish shape, contain a lot of air lacunae in tissues.

■ *In a dry environment*: cacti.

The leaves store a reserve of water.

■ *In a cold environment*: alpine plants.

The surface of the leaves is reduced.

b) Adaptation for various functions

Supportive role: transformation of leaves to tendrils: the pea (*Pisum sativum*), leaves to adhesive discs: the Virginia creeper (*Ampelopsis veitchii*).

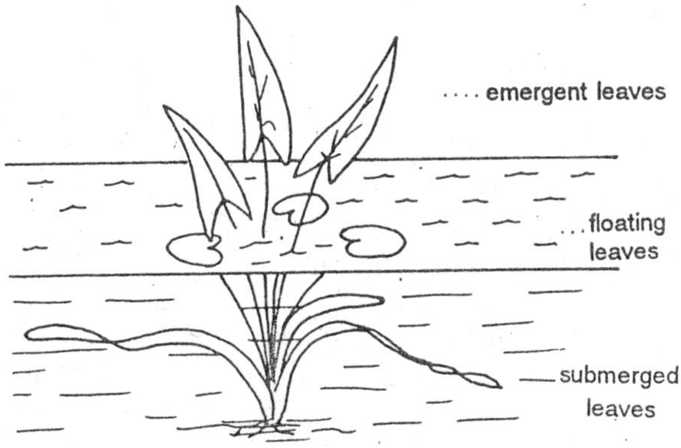

Fig. 2.44. Adaptation of the leaves of the arrowhead to an aquatic environment

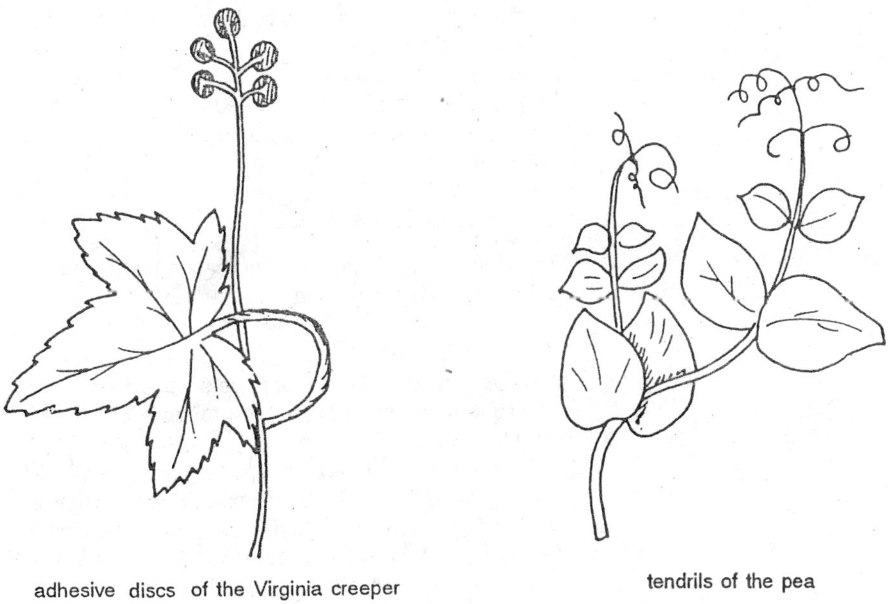

Fig. 2.45. Adaptation of leaves for a supportive role

■ *Protective role*: transformation of leaves into thorns—holly (*Ilex aquifolium*)—a protection that counterbalances the slow growth of such plants *vis-à-vis* "grazing" animals; leaves in the form of scales to protect the buds.

Fig. 2.46. Protection: thorny leaves of holly

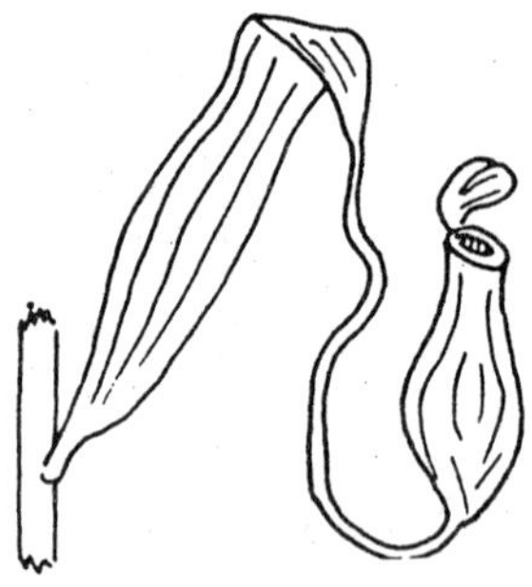

Fig. 2.47. Nourishment: the pitchers of the nepenthes

■ *Role in absorption of nutritive matter*: transformation of leaves into pitchers—nepenthes; these carnivorous plants capture insects in the pitchers and, subsequently, "digest" them.

■ *Role in the accumulation of reserve materials*: scales in bulbs, cotyledons in seeds.

■ *Role in reproduction*: Floral organs are modified leaves (*see* §1.2.4.2).

1.2.3.5 Increase in leaf size

The extent of a leaf surface is increased first at the tip, next in the middle, and finally at the base.

1.2.4 The flower

At the beginning of this chapter we selected six plants to observe, on each of them, the presence of different parts that formed the flower.

All these six were of different shapes and sizes.

We will now examine a very simple, complete flower that is often seen in gardens: the yellow wallflower (*Cheiranthus*) that appears from April till June. Let us pluck one of these flowers.

1.2.4.1 Different parts of the flower

With a pair of small forceps let us separately remove all the organs of the flower that are supported by a *pedicel* inserted on a stem and ending in a broadened part: *the receptacle*.

We will find four types of organs as we move inwards from the outside.

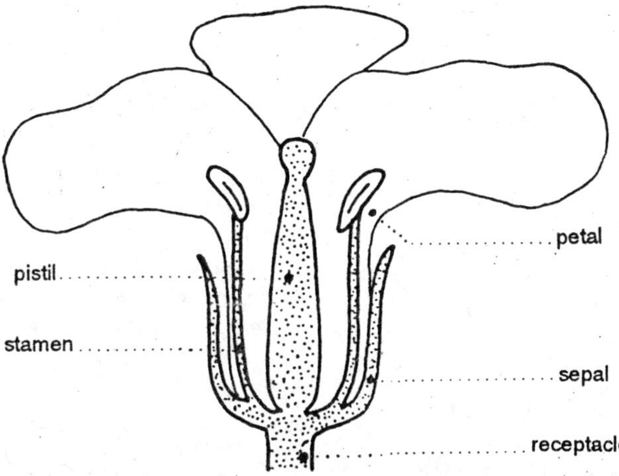

Fig. 2.48. Parts of the flower

a) Four small green blades: the four *sepals* that form the *calyx*.
b) Four parts, which are larger, of a yellow colour arranged in the form of a cross (whence the name of crucifers given to such plants) each formed by a flat part ending in a claw; the four *petals* form the *corolla*.
c) Six swellings, each supported by a filament: these are stamens comprising an anther supported by a stalk.
Two of the stamens are smaller than the others.
All the stamens together form the *androecium*.
d) An elongated, bottle-shaped, narrow organ with a short neck is the pistil in which the swollen portion is the ovary, the extended neck the style and the viscous tip the stigma.
When we open the pistil we observe two cavities each containing developing seeds: the ovules.
We say that the pistil of the wallflower has two carpels.
The pistil is also known as the *gynoecium*.
It may be noted that all floral parts are separate from each other and with respect to one another.

1.2.4.2 Role of different parts of the flower

The calyx and the corolla together form the perianth, which is designed to protect the reproductive organs.
The androecium includes all the male organs and the gynoecium includes all the female organs of the flower.[23]
Their róle ends in the sexual reproduction of flowering

23
In all horticultural plantations that are required to produce fruits or seeds, it is necessary to encourage the formation and activity of the reproductive organs (androecium + gynoecium).

plants (*see* §1 of *Chapter* 6), which then continues with fruiting and the formation of seeds.

1.2.4.3 Different types of flowers

Let us now examine different floral organs and observe their modification with respect to the wallflower (*Cheiranthus cheiri*).

a) The pedicel

If one does not exist, *the flower is sessile*: Aphelandra. At its base there may be a somewhat modified leaf, *the bract*: poinsettia (*Euphorbia pulcherrima*).

Fig. 2.49. A flower and its single bract

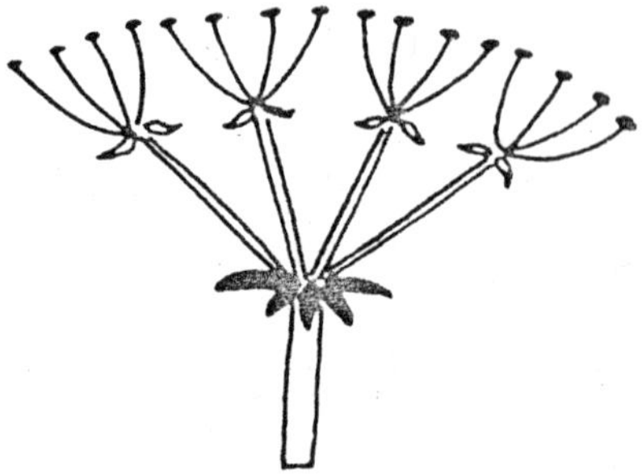

Fig. 2.50. The inflorescence and its bracts form the involucre

If there are several bracts on a flower they together are called an involucre: carrot (*Daucus carota*)

b) The calyx

If the sepals are coloured, the calyx is petaloid: columbine (*Aquilegia coerulea*), hyacinth (*Hyacinthus*), tulip (*Tulipa gesneriana*).

If the sepals are free of each other the flower is dialysepalous or polypetalous: poppy (*Papaver nudicaule*), wallflower (*Cheiranthus cheiri*).

If the sepals are partly united the flower is gamosepalous: carnation (*Dianthus caryophyllus*), digitalis (*Digitalis purpurea*), campanula (*Campanula carpatica*).

The calyx may be deciduous; it most frequently falls off with the corolla.

The *calyx* may also be persistent and is then seen on the fruit: apple (*Malus communis*).

The calyx may also be covered with a second whorl of smaller sepals, this is a calicule or epicalyx: strawberry (*Fragaria vesca*).

c) The corolla

If the petals are free of one another *the flower is dialypetalous or polypetalous*: wallflower (*Cheiranthus

cheiri), violet (*Viola cornuta*), strawberry (*Fragaria vesca*), cabbage (*Brassica oleracea*), dog rose (*Rosa canina*).
In this case the petal is formed by a blade and claw.
If the petals are united, *the flower is gamopetalous*: campanula (*Campanula carpatica*), tobacco (*Nicotiana tabacum*), primrose (*Primula hortensis*), digitalis (*Digitalis purpurea*), potato (*Solanum tuberosum*).
The corolla then forms a tube that widens somewhat at its opening.

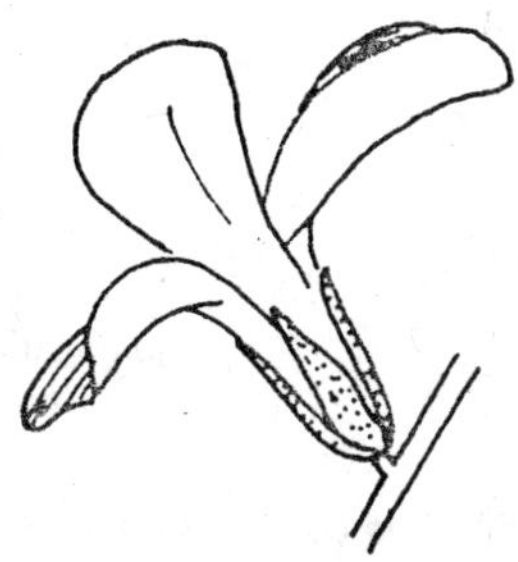

dialysepalous flower of the wallflower

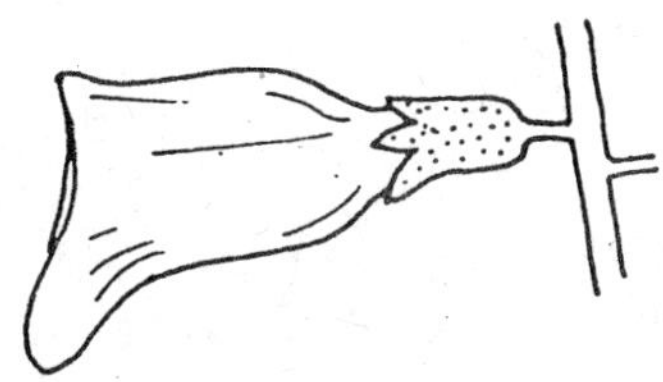

gamosepalous flower of the digitalis

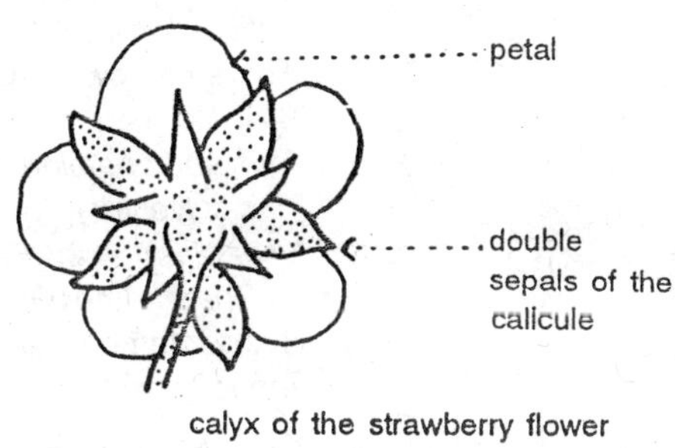

calyx of the strawberry flower

Fig. 2.51. Calyx of the flower

If there are no petals *the flower is apetalous*: anemone of the florists (*Anemone coronaria*), blue clematis (*Clematis viticella*), elm (*Ulmus americana*), beech (*Fagus sylvatica*).
What remain are petaloid sepals.
Generally, on flowers we find:
— in dicotyledons: four to five petals,
— in monocotyledons: three or multiples of three petals.
Notes on the perianth (calyx + corolla)[24]
1) If the sepals and petals are equal, *the flower is regular*: wallflower (*Cheiranthus cheiri*), strawberry (*Fragaria vesca*).

24
Various shapes and colours of the perianth in flowers are an asset for decorative purposes in gardens.

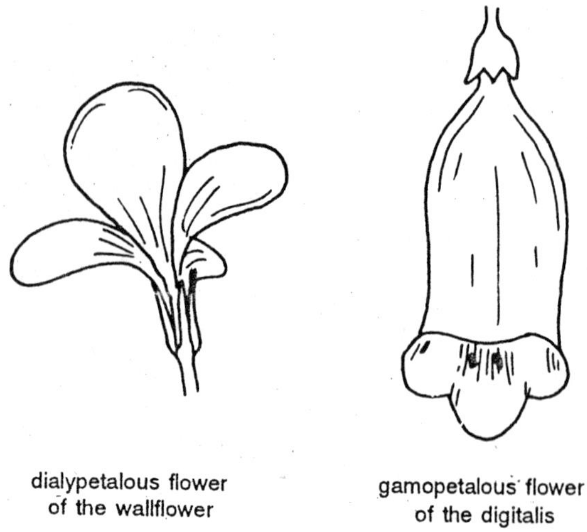

Fig. 2.52. The corolla of a flower

If they are not equal, the flower is irregular: pea (*Pisum sativum*), bean (*Phaseolus vulgaris*).
There is, however, a bilateral symmetry.
2) If there are no sepals or petals, *the flower is asepalous and apetalous, it is aperianthous*: Scotch pine (*Pinus sylvestris*), willow (*Salix babylonica*), poplar (*Populus nigra*), ash (*Fraxinus excelsior*).
d) The androecium
If the stamens are separate, *the flower is dialystemonous*: wallflower (*Cheiranthus cheiri*).

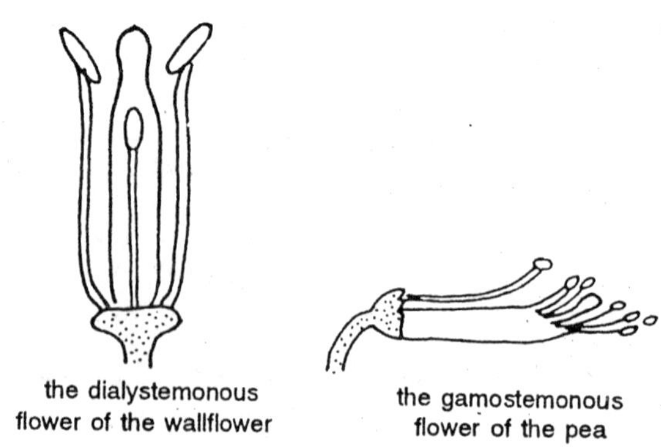

Fig. 2.53. The androecium of a flower

If the stamens are united, *the flower is gamostemonous.*

This union can be through filaments—pea (*Pisum sativum*)—or through the anthers—Compositae.

If there are no stamens, *the flower is astemonous*: it is a female flower and is *unisexual.*

The filaments may be of the same or of different lengths: wallflower (*Cheiranthus cheiri*).

e) The gynoecium or pistil

Depending upon the plant, we observe, per pistil:

- one carpel: pea (*Pisum sativum*), bean (*Phaseolus vulgaris*);
- two carpels: radish (*Raphanus sativus*); or
- several carpels: apple (*Malus communis*), strawberry (*Fragaria vesca*).

If the carpels are separate, *the flower is dialycarpous/apocarpous*: garden ranunculus (*Ranunculus asiaticus*), strawberry (*Fragaria vesca*).

If the carpels are united, *the flower is syncarpous*; this is the most frequent case: cabbage (*Brassica oleracea*), wallflower (*Cheiranthus cheiri*), apple (*Malus communis*).

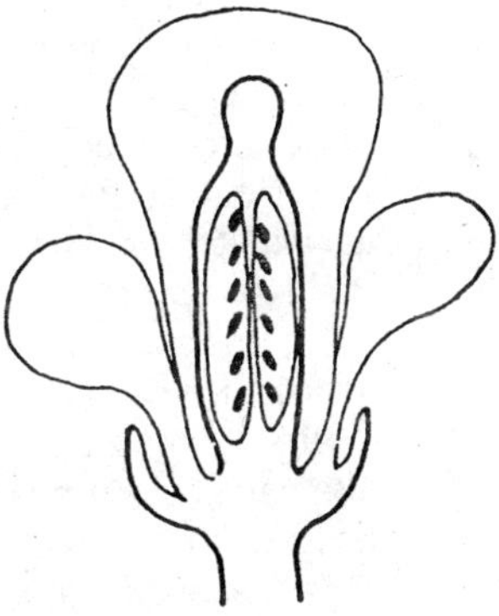

syncarpous flower of the wallflower

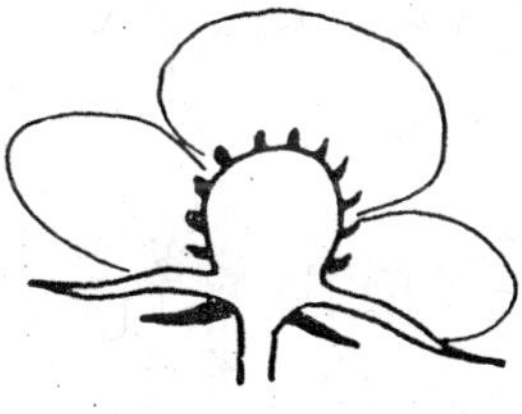

dialy carpous flowers of the strawberry

Fig. 2.54. The pistil of a flower: position of the carpels

If there are no carpels, *the flower is acarpous.* It is then a male flower and is unisexual.

The carpels are attached to the receptacle, which may be flat or in the shape of a somewhat deep cup.

If the ovary is inserted on a flat receptacle and is, therefore, free of the other floral parts, we say that *the ovary is superior*—for example, the poppy (*Papaver somniferum*); the carpels may then be free.

In the opposite case, *the ovary is inferior*; for example, the vegetable marrow (*Cucurbita pepo*); the carpels are always united.

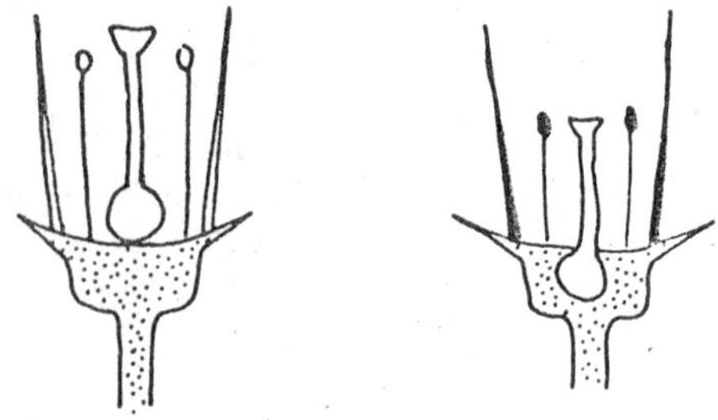

Fig. 2.55. The pistil of a flower: position of the carpels

We make a transverse section of a pistil and can, thereby, see (most often with a magnifying glass):

— the number of carpels,

— whether they are free or united,

— the manner in which the ovules are attached within the ovary, especially if the carpels are united.

This last characteristic is called the *placentation*. If the carpels are united along their walls, we observe as many loculi as there are carpels and the ovules are attached on the axis of the pistil; *the placentation is axile*: lilac (*Lillium candidum*).

If the carpels form only one loculus and the ovules are attached to the walls of the pistil *the placentation is parietal*: violet (*Viola cornuta*).

If the carpels form a single loculus and the ovules are attached on a central axis, *the placentation is central* (a somewhat rare occurrence): primrose (*Primula hortensis*).

Note on the pistil of gymnosperms

The existence, or absence, of an ovary in the carpel determines a classification of phanerogams, or seed plants.

The composition of the gynoecium, as we have seen, applies to the sub-branch angiosperms, a term that signifies: ovule within a "sac".

In gymnosperms, such as Scotch pine (*Pinus sylvestris*), we can say that the carpels do not close up to form this "sac" but remain spread out. The ovules are placed on them; they are not covered. The ovules are naked.

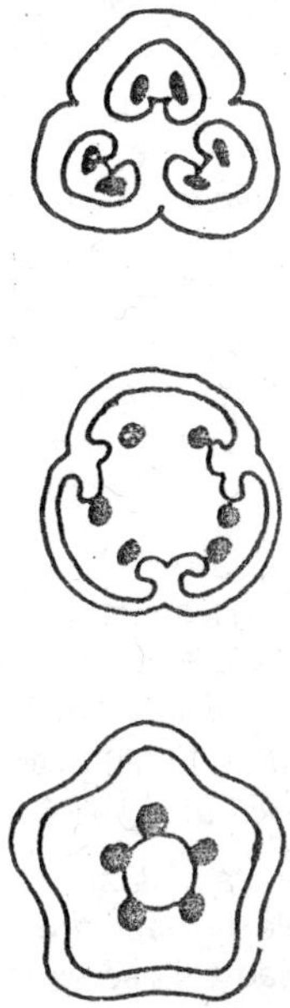

axile placentation
of the lily

parietal placentation
of the violet

central placentation
of the primrose

Fig. 2.56. Arrangement of ovules in a flower

In the female cone, between every two bracts of the cone, we find carpellary scales on which the ovules are fixed.

It may be noted that if there is no ovary there are no fruits and fertilised ovules result in seeds that, when ripe, escape directly from the female cone.

Note on unisexual flowers: we know that the androecium and the pistil constitute the reproductive organs of a flower.

In most cases they are both formed on the same flower. *Such a flower is hermaphrodite.*

Sometimes we come across acarpous or astemonous flowers. Nevertheless, female and male flowers may be found on the same plant; *such plants are monoecious*: oak (*Quercus sessiliflora*), melon (*Cucumis melo*). If they are found on two different plants, *the plants are dioecious* and we then have male and female stocks: willow (*Salix babylonica*), date palm (*Phoenix dactylifera*).[25]

25

If it is intended to obtain the fruit of a species it is obvious that we must cultivate plants of both sexes or ensure that female flowers are pollinated.

1.2.4.4 Inflorescences

In our initial examples we had examined the single flower of the tulip. On the other hand, the myosotis

(*Myosotis alpestris*), the French marigold (*Tagetes patula*), the apple (*Malus pumila*), etc. have flowers that are combined on the same floral axis. The combination of these flowers is called an inflorescence.

a) Indefinite inflorescences

An inflorescence is indefinite when the axis:

— ends in a vegetative bud and never in a terminal flower,
— may elongate indefinitely,
— and bear flowers laterally.

There are several such types:

■ *The spike*

The axis, bears an indeterminate number of sessile flowers laterally: willow (*Salix babylonica*), oak (*Quercus sessiliflora*), ribgrass (*Plantago lanceolata*).
Flowering begins at the bottom of the spike.
The youngest flowers are towards the tip.

■ *The capitulum*

Sessile flowers are attached to the upper end of a widened axis so as to form a flat surface that is encircled by a number of small modified leaves or bracts. These bracts form the involucre: daisy (*Leucanthemum maximum*), artichoke (*Cynara scolymus*), dahlia (*Dahlia hortensis*).
Flowering begins at the periphery of the capitulum.
The youngest flowers are in the centre.

■ *The raceme*

This resembles a spike but the flowers are pedicelled and the secondary axes are of the same length: wallflower (*Cheiranthus cheiri*), bean (*Phaseolus vulgaris*), cytisus (*Cytisus laburnum*), red currant (*Ribes rubrum*), foxglove (*Digitalis purpurea*).
When the secondary axes are longer in the centre than at the two ends of the cluster, the inflorescence is a thyrse: horse chestnut (*Aesculus hippocastanum*).
Flowering begins at the bottom of the cluster.
The youngest flowers are at the end.

■ *The umbel*

The secondary axes are all of the same length but begin from the same point. Each of them ends in a flower.
The bracts of each of the floral axes are joined at the same point and form the involucre: primrose (*Primula hortensis*), cherry (*Cerasus avium*), onion (*Allium cepa*).
Flowering begins at the periphery of the umbel.
The youngest flowers are in the centre.

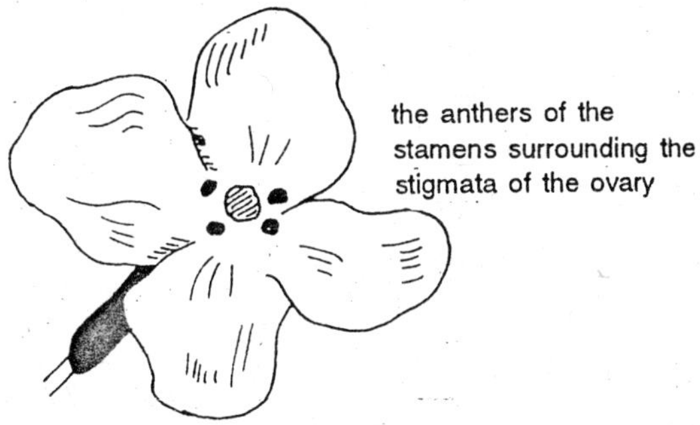

hermaphrodite flower

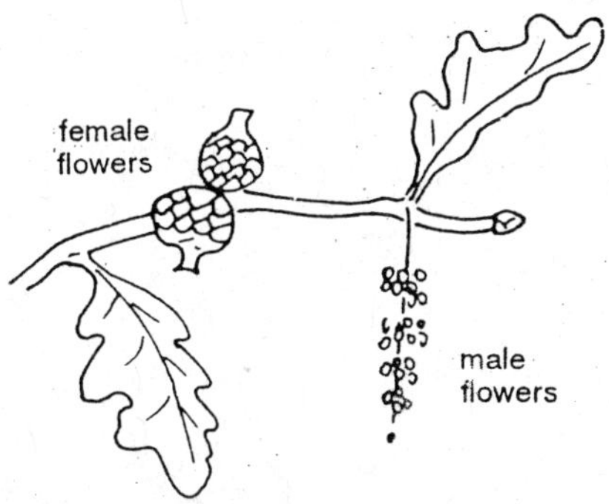

monoecious plant: oak

dioecious plant: willow

Fig. 2.57. Hermaphrodite and unisexual flowers

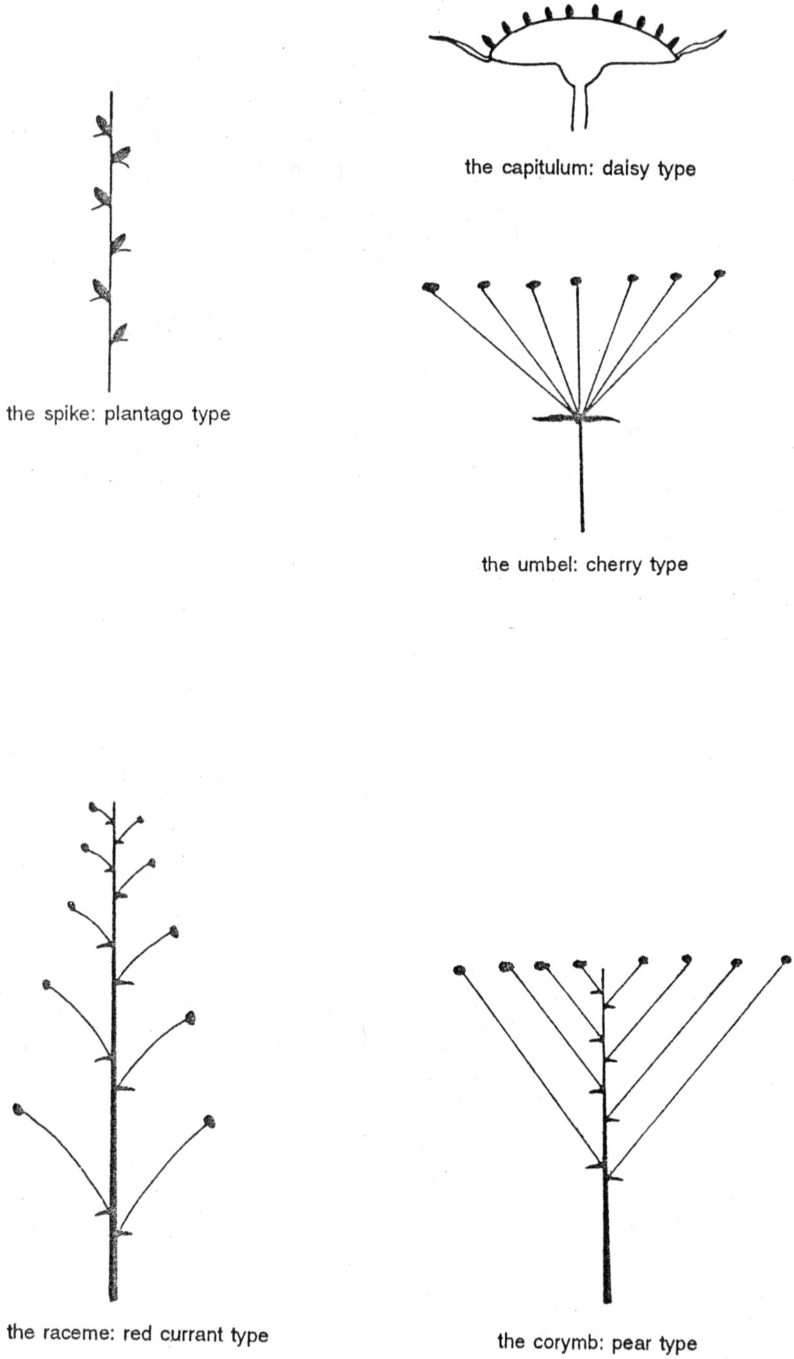

Fig. 2.58. Indefinite inflorescences

■ *The corymb*

This resembles a raceme but all the secondary axes end at the same height: pear (*Pyrus communis*), hawthorn (*Crataegus oxyacantha*), plum (*Prunus domestica*).

Flowering begins at the periphery of the corymb.

The youngest flowers are in the centre.

b) Definite inflorescences

These are also called *cymes.*

The primary axis ends in a flower and, at its base, has one or several bracts. From this point one or two secondary axes begin that behave as the primary axis, and so on.

Instead of increasing in length the inflorescence increases in breadth.

Flowering begins with the flower on the primary axis and then the one on the secondary axis, etc.

We can identify:

■ *Uniparous cymes*

A single axis begins at each bifurcation:

— Always on the same side, *scorpioid uniparous cyme*: myosotis (*Myosotis alpestris*), comfrey (*Symphytum officinale*).

— Alternately from one side and then from the other, *uniparous helicoidal cyme*: gladiolus (*Gladiolus*), day lily (*Hemerocallis flava*).

■ *Biparous cymes*

Two axes begin at each bifurcation: gypsophila (*Gypsophila paniculata*).

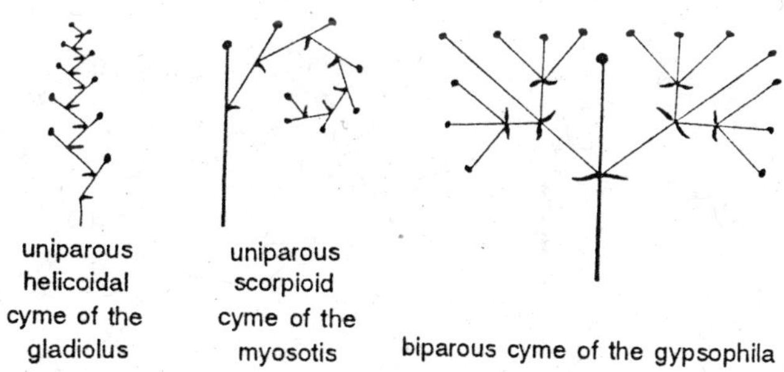

Fig. 2.59. Definite inflorescences

c) Compound inflorescences

In certain cases the inflorescence is formed normally but each flower is replaced with a secondary inflorescence.

the panicle of spikelets of the oat

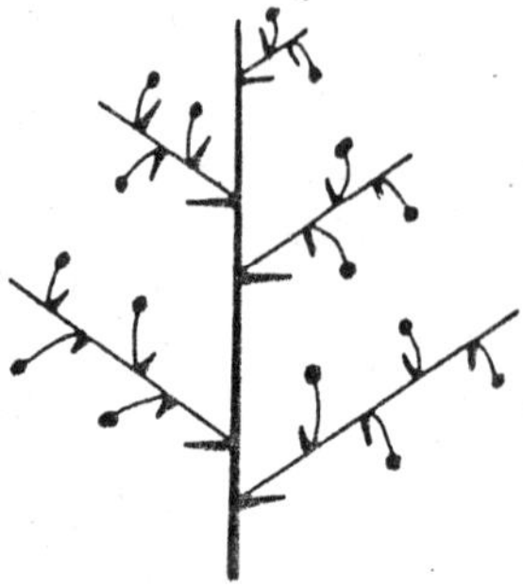

the panicle of racemes of the lilac

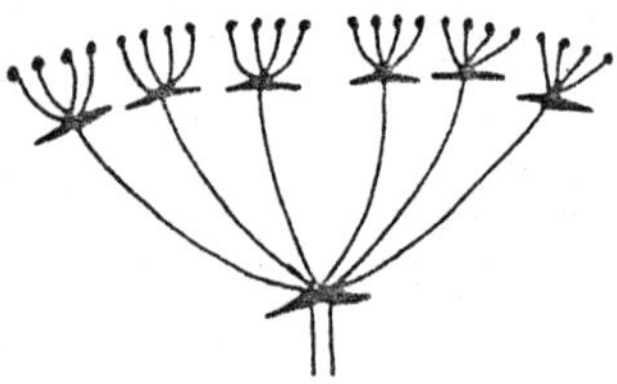

the umbel of umbels of the carrot

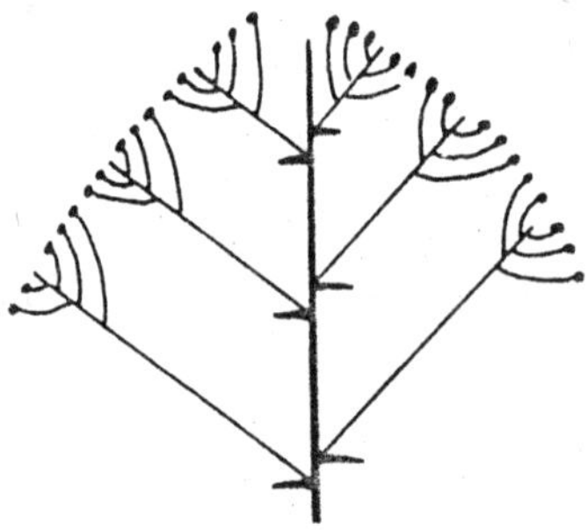

the corymb of corymbs of the service tree

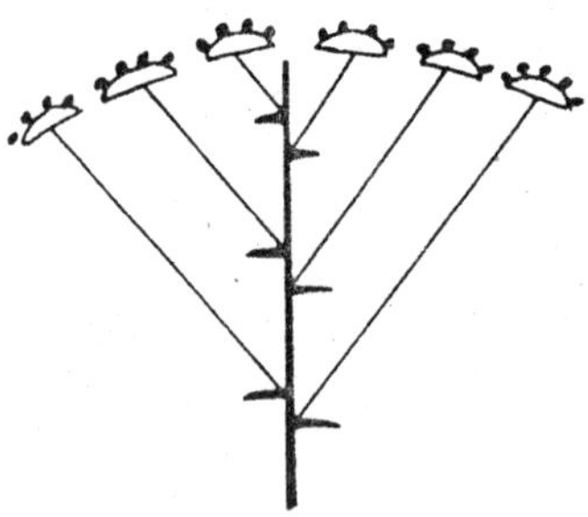

the corymb of capitula
of the milfoil yarrow

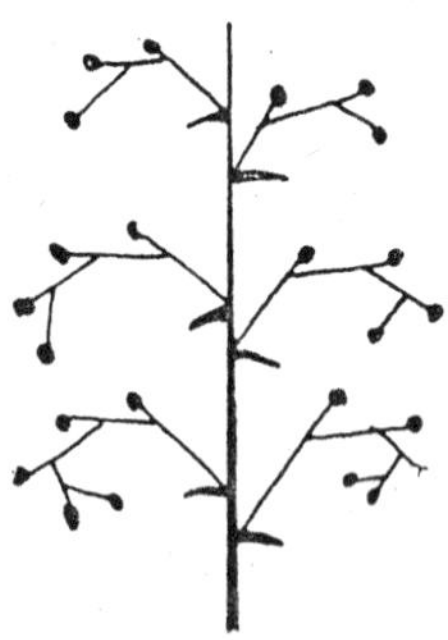

the panicle of scorpioid
cymes of the seringa

Fig. 2.60. Compound inflorescence

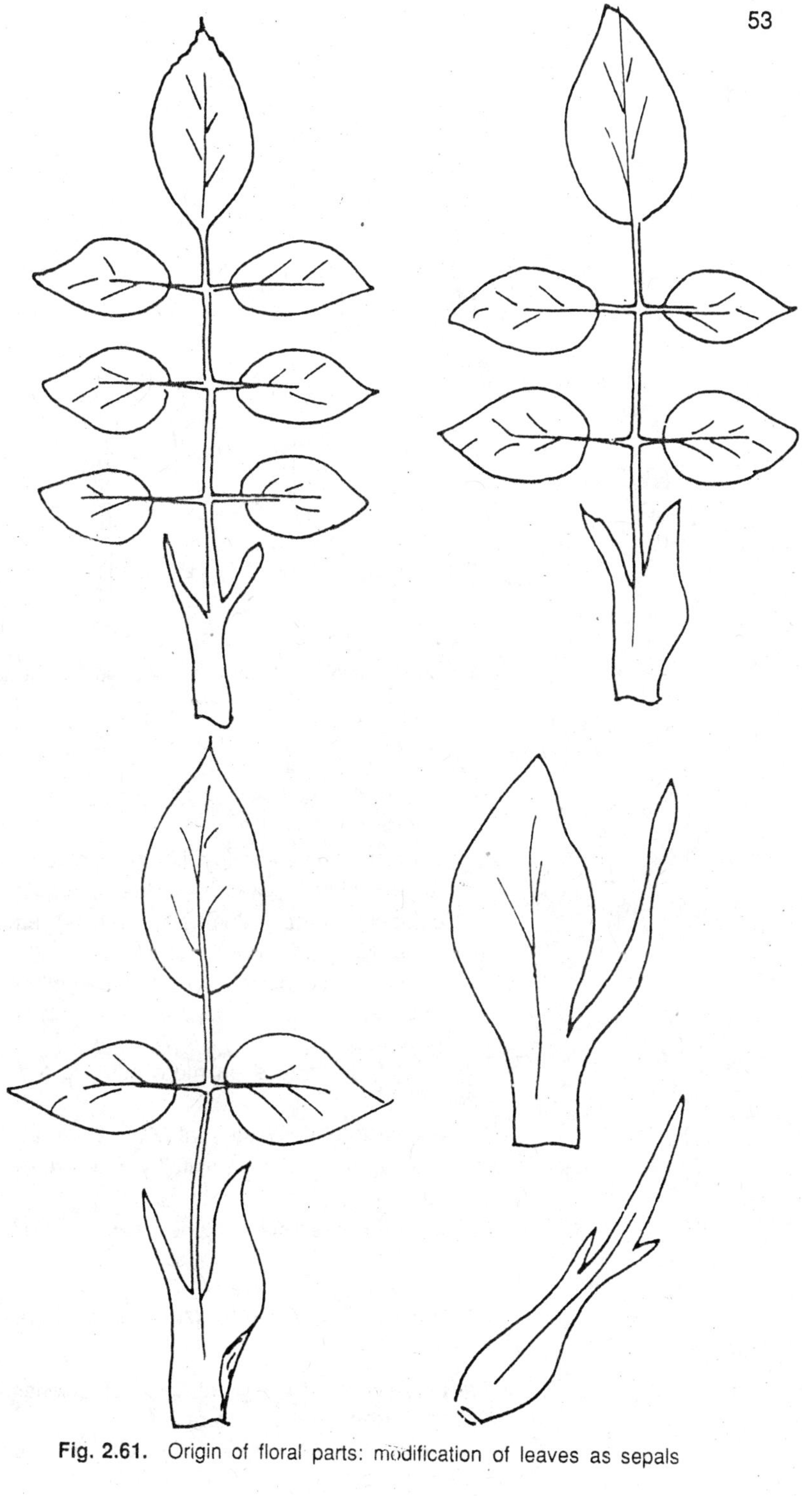

Fig. 2.61. Origin of floral parts: modification of leaves as sepals

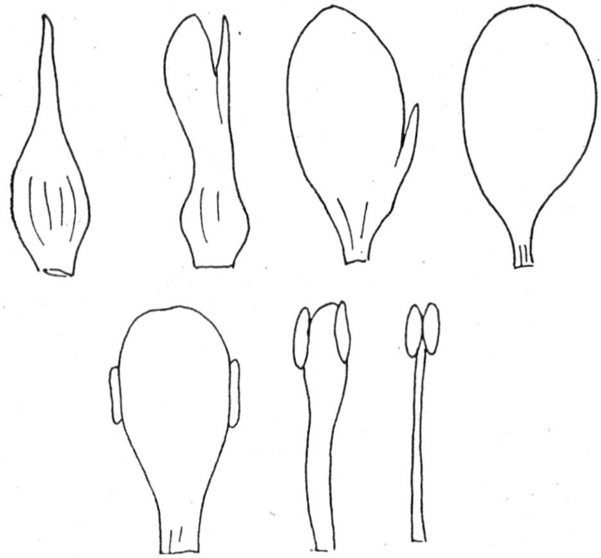

Fig. 2.62. Origin of floral parts: modification of sepals as stamens

We thus encounter:

— compound spikes (spike of spikelets): wheat (*Triticum sativum*);

— compound racemes (panicle of racemes): lilac (*Syringa vulgaris*), privet (*Ligustrum ovalifolium*);

— compound umbels (umbel of umbels), umbellifers, carrot (*Daucus carota*);

— compound corymbs (corymb of corymbs): the ash (*Sorbus latifolia*), European elder (*Sambucus nigra*), etc.

We may also have even more complicated associations:

— a panicle of spikelets: oat (*Avena sativa*),

— a corymb of capitula: milfoil yarrow (*Achillea millefolium*),

— a panicle of scorpioid cymes: seringa (*Philadelphus coronarius*).

1.2.4.5 Origin of floral parts. Floral parts are modified leaves. Proof of this lies in:

a) the definite arrangement, for each species, of floral parts as seen in the case of leaves;

b) the existence of venations in the limbs of leaves that are also seen in floral parts;

c) the identical appearance of the floral bud and the leaf bud.
d) the stages between leaves and sepals—hellebore (*Helleborus abschasicus*), peony (*Paeonia lactiflora*); between the sepals and other parts—white water lily (*Nymphaea alba*).[26]

1.2.4.6 Flower formation[27]

Natural floral development may be altered by various substances, some inducing flowering and others slowing down the growth of flowers, and even their fading.

1.2.5 The fruit

We will study the phenomena of fruiting in the context of *sexual reproduction* of flowering plants.
We know that:
— fertilised ovules produce seeds,
— the ovary develops as a fruit,
— petals and stamens fade,
— sepals are visible on the fruit if they are persistent,
— the floral receptacle becomes the receptacle of the fruit,
— the floral pedicel becomes the pedicel of the fruit.
This is what usually happens.

1.2.5.1 Various parts of a fruit

Let us examine a sectioned fruit of the plum tree (*Prunus domestica*) cultivar (*see definition in* §*1.1.2, Chapter 8*) Reine Claude Dorée.
Moving inwards from the outside we observe:
a) a skin, a type of thick yellow film,
b) a yellowish fleshy part that is sweet and juicy,
c) a very hard "stone" containing
d) a seed: the "almond".
We have already mentioned that a fruit emanates from the development of the ovary and contains a seed or seeds.
We may, therefore, conclude that this "almond" is the seed of the plum tree and the three other parts emanate from a development of the ovary.
We are aware that the ovary is a modified leaf as are all other floral parts and is formed by its almost complete folding.
From this we deduce that:
a) The skin is the external epidermis of this carpellary leaf; it is the epicarp.
b) The fleshy portion is its parenchyma; it is the mesocarp.

26
a. Breeders are able to transform the stamens of certain flowers to petals: rose (*Rosa*), peony (*Paeonia lactiflora*), ranunculus (*Ranunculus asiaticus*).
The almost total disappearance of reproductive organs, however, prevents sexual multiplication even if the variety is an established one.
b. General use of flowers:
i) Obviously, in floriculture.
ii) As food products:
— capers are the floral buttons of the caper plant macerated in vinegar.
— cloves are the dried floral buds of the clove tree.
— the cauliflower is an enlarged inflorescence.
iii) As medicinal plants:
—principally against colds: red poppy, mallow, violet, marshmallow.
iv) As plants for perfumes:
—rose, jasmine, violet, hyacinth, carnation, mimosa, bitter orange, etc.

27
The horticulturist may reduce the period before flowering of Bromeliaceae. For example, *etefon* (X.TR.4), which, by a breakdown in the plant, after a systematic circulation of the active product, is transformed into ethylene and induces flowering.
On the other hand, a horticulturist can slow down the fall of flowers. For example, silver sulphate often suppresses the "stress" involved in transportation of flowering plants. Untreated flowers may remain fresh for a few days but with treatment this period may extend to two or three weeks.

c) The stone is its internal epidermis; it is the endocarp but, in the plum, it has become very hard; it is lignified. These three parts, together, are known as the pericarp.

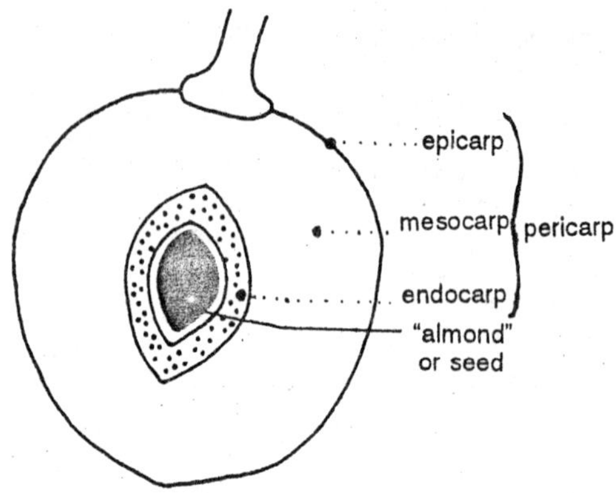

Fig. 2.63. Different parts of a fruit

The mode of development of the pericarp enables the classification of fruits.

1.2.5.2 Different types of fruits

When the pericarp is thick, soft and fleshy, the fruit is fleshy. When the pericarp is dry and thin, the fruit is dry.

a) Fleshy fruits

Fleshy fruits do not release their seeds spontaneously. We call them indehiscent fruits.

There are three types of fleshy fruits.

■ *Berries*

For example, red currants (*Ribes rubrum*), grapes (*Vitis vinifera*).

A "seed" of these fruits has a skin, pulp and seeds—the pips.

The non-lignified endocarp is, therefore, a part of the pulp because we cannot identify it separately.

To summarise: *berries* have an epicarp: a skin, a fleshy mesocarp and endocarp, as well as seeds.

□ *Specific berries*

The date (*Phoenix dactylifera*) is a berry with a lignified seed.

The orange (*Citrus sinensis*) and the lemon (*Citrus limon*) are composite berries with 6 to 12 carpels joined together. Hairs from the endocarp are swollen

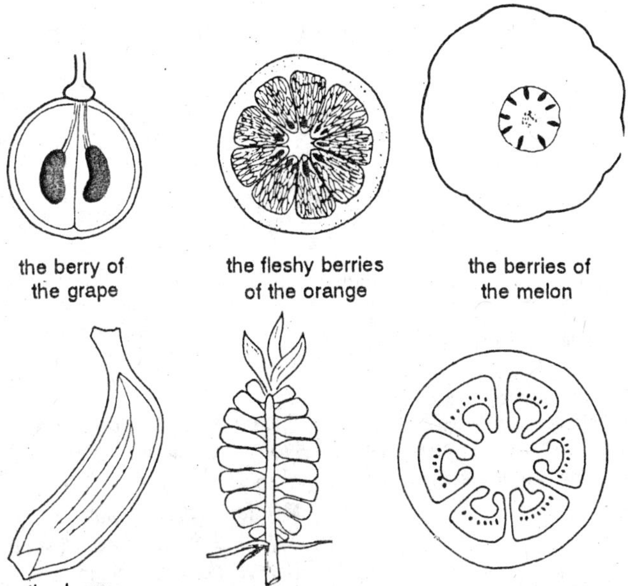

Fig. 2.64. Different types of berries

with juice and form the pulp. The epicarp and the mesocarp form the outer skin, called the zest.

The melon (*Cucurbita melo*) and the pumpkin (*Cucurbita pepo*) are enormous berries on trailing stems.

The banana (*Musa ensete*) is a berry in which the seeds are underdeveloped.

The pineapple (*Ananas comosus*), somewhat more complicated, is formed of several agglomerated berries and bracts.

The tomato (*Solanum lycopersicum*) is a berry in which the seeds are surrounded by a fairly large amount of jelly.[28]

■ *Drupes*

For example, plums (*Prunus domestica*), peaches (*Prunus persica*), olives (*Olea europaea*), which we have studied and in which the lignified endocarp constitutes the stone.

□ *Special features of drupes*

Of the walnut (*Juglans regia*) only the seed is edible.

The raspberry (*Rubus idaeus*) and the mulberry (*Morus alba*) are made up of many drupes attached to the receptacle of the fruit.

The coconut (*Cocos nucifera*) is the stone of a drupe.

28

a. The horticulturist will try to select plants with berries that:
— are largest,
— have a flesh that is very sweet and juicy (in the case of fruits).

b. For trials on seedless fruits by hormone treatment, see Chapter 5, Physiology.

c. For the tomato, we attempt to reduce the volume of jelly and, especially, to make it as little liquid as possible.

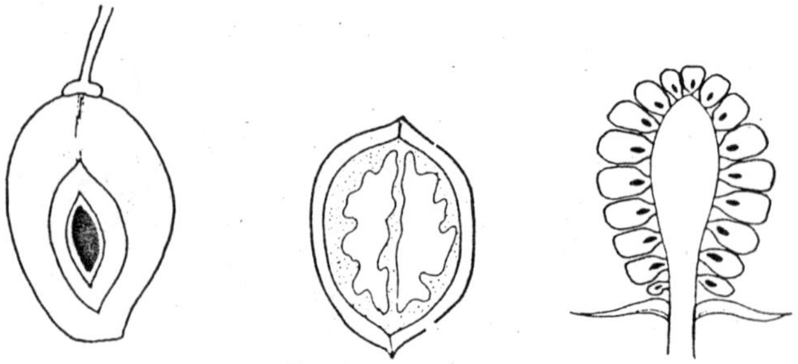

Fig. 2.65. Different types of drupes

It is composed of three parts: a solid endosperm—"the ivory", a liquid endosperm—the "milk", and an embryo.[29]

■ *Intermediate fruits*

For example, apple (*Malus communis*) and pear (*Pyrus communis*), in which the endocarp is neither fleshy (berry) nor lignified (drupe); it is cartilaginous. Hence, these fruits are called intermediate fruits or pomes.[30]

The mass becomes fleshy by a thickening of the base of the sepals, the petals, the stamens or the receptacle and is attached to the wall of the ovary.

29

For drupes the arboriculturist wants cultivars in which the edible portion is much larger than the stone: endocarp + almond.
This, of course, is applicable only to cases where the fleshy portion of the drupe is edible.

30

This taste value of the fruit, in this case, is most important.

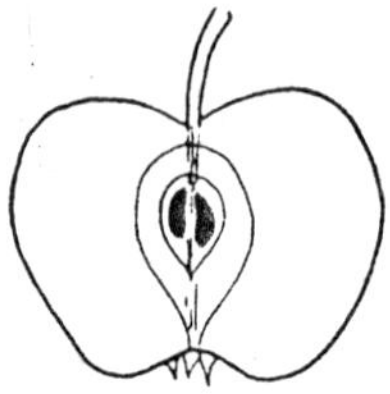

Fig. 2.66. Pomes: the apple

b) Dry fruits

They are dehiscent or indehiscent.

■ Dry indehiscent fruits: The seeds are not released by the fruits. These are *achenes*, fruits with a single seed: oak (*Quercus pedunculata*), hazel (*Corylus avellana*), ranunculus (*Ranunculus asiaticus*). In the Cupuliferae, the achene is surrounded by an involucre: *the cupule.* Achenes may be:

— Double, diachenes: carrot (*Daucus carota*).

— Numerous, polyachenes: strawberry (*Fragaria vesca*); the edible portion is the receptacle that has become fleshy.

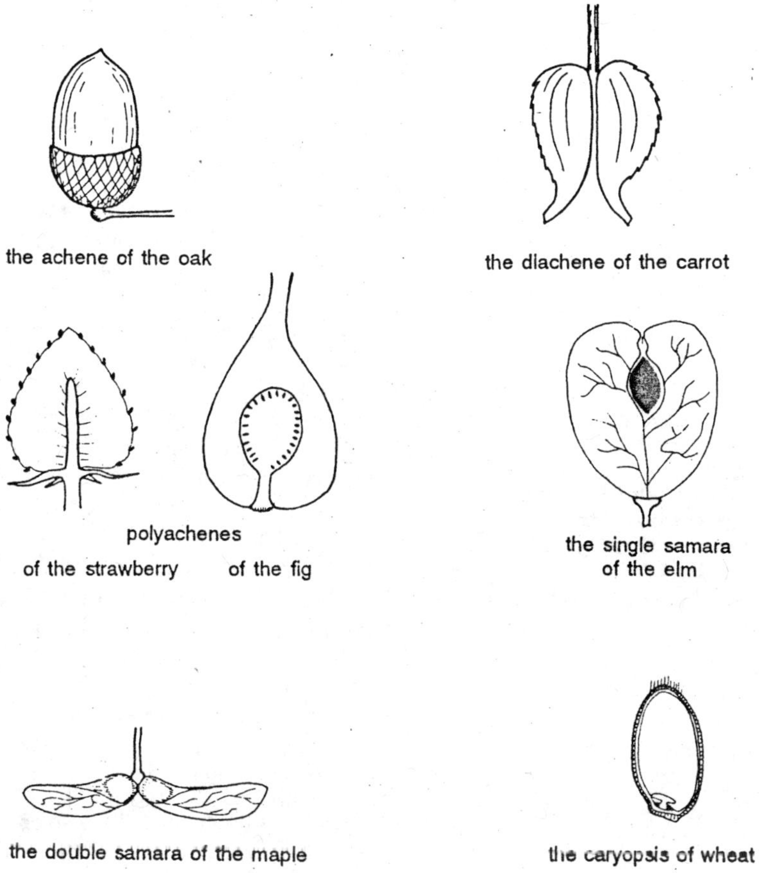

Fig. 2.67. Dry indehiscent fruits

Achenes are seeds found on the "flesh" of a strawberry or fig (*Ficus çarica*); the inflorescence, having become fleshy and hollow, supports many achenes.
The fruits of the artichoke (*Cynara scolymus*) are achenes. We consume the fleshy base of the bracts on the flower as well as the receptacle. The "choke" is formed by flowers that have not blossomed.

— Winged, *samaras* with one wing—elm (*Ulmus montana*)—or with two wings—maple (*Acer campestre*).

If the pericarp is attached to the seed it is a *caryopsis*, a fruit with a single seed: wheat (*Triticum sativum*).

■ Dry dehiscent fruits: The seeds are released from the fruits.

Depending on their shapes and especially on the number of carpels, we may have:

— *follicles*: fruits with a single carpel that opens along a longitudinal slit at the joint of the placentae: hellebore (*Helleborus abschasicus*) var. *"colchicus"*), peony (three connected follicles) (*Paeonia lactiflora*);
— *pods*: fruits with a single carpel that opens out through two oppositely placed longitudinal slits. They are also called "legumes" (leguminous plants);
— *siliquas*: fruits with two carpels. The central partition is fixed; the two valves open through four slits and remain attached to the top. The seeds are fixed to the partition: cabbage (*Brassica oleracea*), wallflower (*Cheiranthus cheiri*). Sometimes a siliqua is constricted at each seed: radish (*Raphanus sativus*).
— *capsules*: fruits with several carpels opening through a circular slit that forms a cover—pimpernel (*Anagallis arvensis*)—or through pores: poppy (*Papaver somniferum*).[31]

31

Many dry fruits are decorative but their importance in horticulture as food plants is much less than that of fleshy fruits. In fact, the only part of a "dry fruit" that can be eaten is its edible seed (or seeds).

1.2.6 The seed

A seed originates from a fertilised ovule.

1.2.6.1 Parts of a seed

Let us examine a bean seed (*Phaseolus vulgaris*). Its traditional kidney shape is well known, as is also the scar in the concave part where it was attached to the ovule, the hilum.

a) Externally, we observe a somewhat thick skin with a coloration (depending on the variety of bean, it can be white, green, red ...) and with a venation. If we remove this epidermis we find another thinner one. These are the two *integuments*.

b) When we break a seed two parts separate easily: these are the two *cotyledons*.

c) Between the two cotyledons we observe the embryo or plantlet that has three parts: a radicle, a tigellum and a plumule.

1.2.6.2 Role of different parts of a seed

a) Integument

This is the protective covering of a seed. There can be one or two such coverings. For example, there are two in a bean seed.

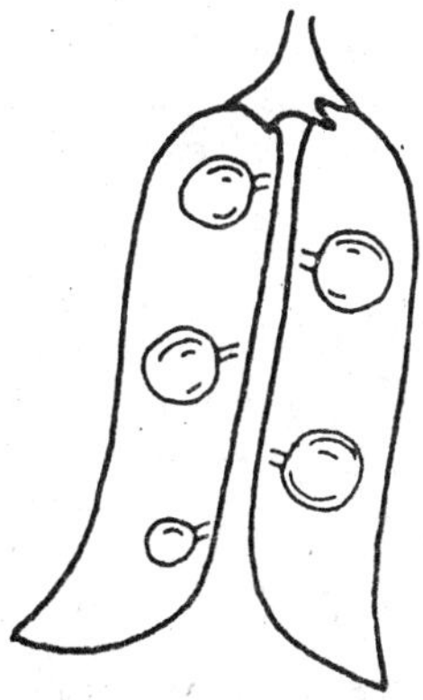

pod of the pea

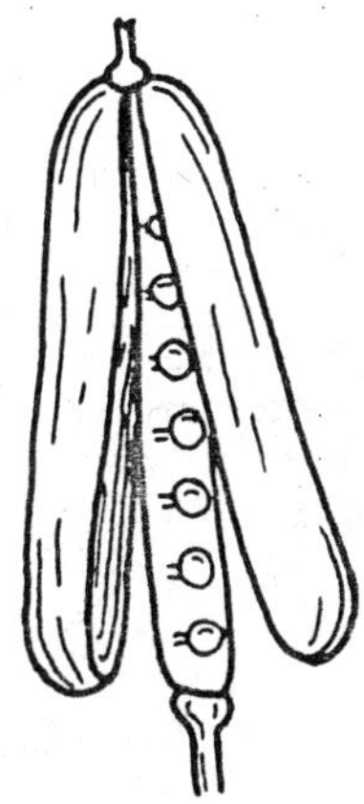

siligua of the cabbage

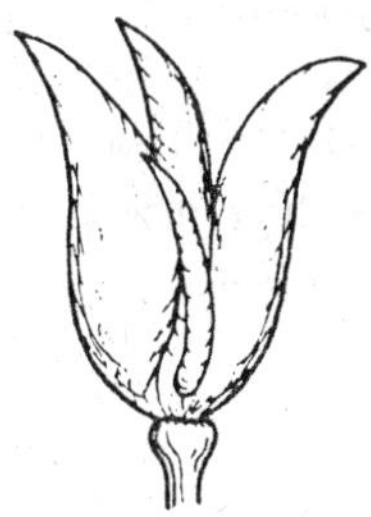

follicles of the peony

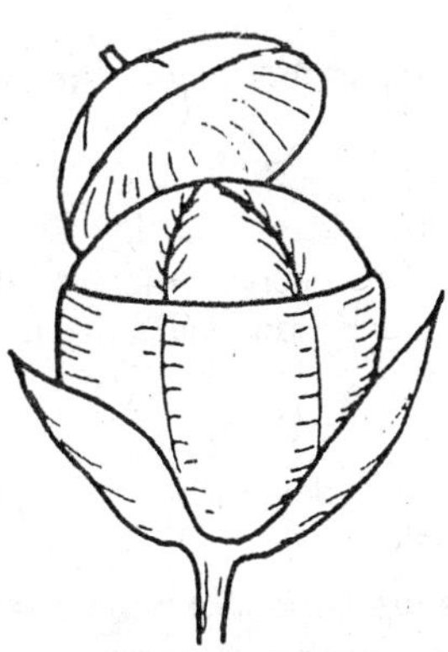

capsule with a cover
of the pimpernel

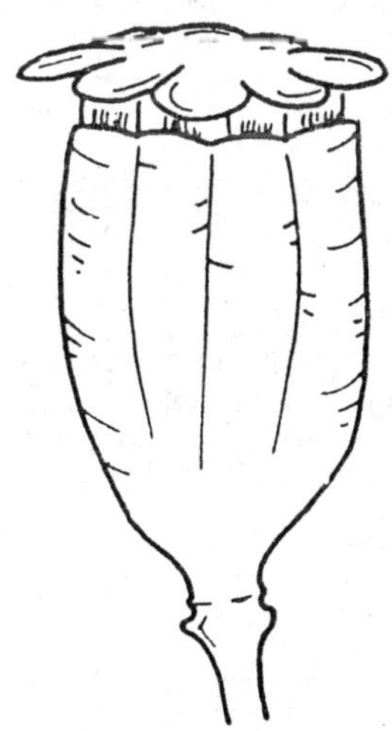

capsule with pores
of the poppy

Fig. 2.68. Dehiscent dry fruits

If we find only one, it is the internal integument that does not exist: the walnut (*Juglans regia*).
This epidermis, more or less impenetrable, depending on the type of seed, enables it to survive until germination.[32]

32

The period for which a seed can be preserved depends on the resistance and impermeability of the integument or integuments.

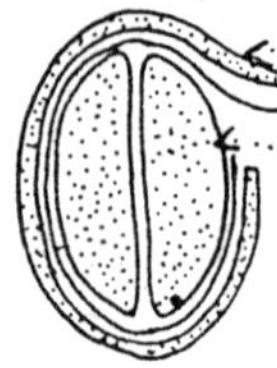

Fig. 2.69. Bean seed (dicotyledons)

b) Embryo or plantlet
This produces the future plant and its three parts correspond to the three principal parts of the plant:
— the radicle provides the future root,
— the tigellum provides the future stem,

Fig. 2.70. Parts of the embryo (or plantlet)

— the plumule provides the future terminal bud and the leaves.

c) The cotyledons
These are, basically, the two first leaves (or leaf) of the plant in a state of reserve organs[33] that ensure the survival of the young plant until it can meet its own requirements.
Two points should be noted:
— There are two cotyledons in a bean seed but we may also find only one.
— The cotyledons are large in a bean seed because it does not contain endosperm or any other reserve tissue.

If a seed contains endosperm, the cotyledons are small and the endosperm surrounds the embryo.

33

The value of the reserves contained in a seed determines the use we can make of it as a food (see § 16, Chapter 5).

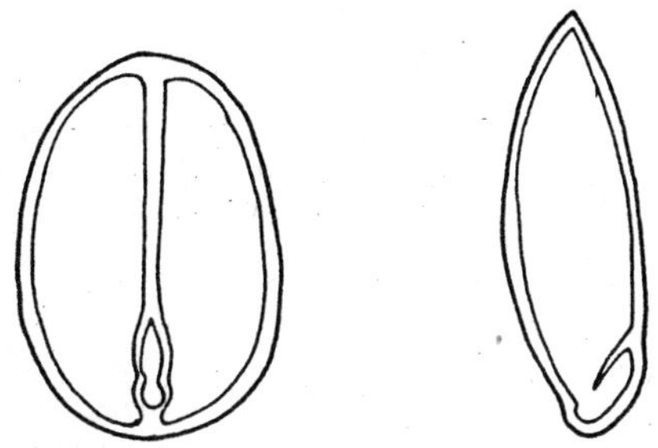

A2 cotyledon (haricot bean)   A1 cotyledon (cereal)

Fig. 2.71. Seed

1.2.6.3 Different types of seeds

The special features of the cotyledons are the basis for classification of seeds.
We observe:

a) Seeds with two cotyledons, whence the class of dicotyledons: pea (*Pisum sativum*), bean (*Phaseolus vulgaris*), castor (*Ricinus communis*);
Seeds with one cotyledon, class of monocotyledons: leek (*Allium porrum*), wheat (*Tricitum sativum*).

b) The endosperm, a reserve tissue, eventually disappears. If this happens before the seed ripens we have a seed without endosperm—pea (*Pisum sativum*), bean (*Phaseolus vulgaris*)—and the reserves will be in the cotyledons.
If it happens after the seed has ripened we have a seed with endosperm—leek (*Allium porrum*), castor (*Ricinus communis*)—with thin cotyledons.

1.3 Modes of growth

We have studied the various morphological parts of flowering plants. This will help us in identifying their different modes of growth.
In fact, as a basis of observation, we have examined plants with different appearances as well as different growth cycles.
We will now classify these cycles of the pine, the apple, the French marigold, wheat, the myosotis, the tulip and the lupin (see the Latin names, given futher on).

1.3.1 Plants that flower only once

Such plants develop their vegetative apparatus, flower once and then die.

In this group we have:
The French marigold (*Tagetes patula*), spring wheat (*Triticum sativum*) and myosotis (*Myosotis alpinus*). All of them, however, do not have cycles of equal duration.
a) Spring wheat and the French marigold flower die during the year in which they germinated.
They are *annual plants*.
b) The myosotis forms its vegetative apparatus during the year of its germination but flowers and then dies during the next year. It is a *biennial plant*.
c) There are, however, certain others such as the century plant (*Agave americana*), an oily plant that fruits long after its germination; several decades after. It is a *pluriannual plant*.

1.3.2 Plants that flower several times

These plants develop their vegetative apparatus, flower several times and then die.
In this group we have the pine (*Pinus sylvestris*), the apple (*Malus communis*), tulip (*Tuiipa X*), hyacinth (*Hyacinthus orientalis*), and lupin (*Lupinus polyphyllus*).
The vegetative unit of each of these does not, however, have the same longevity.
a) A tulip (*Tulipa*) bulb when planted will, each year, yield a flower (progressively smaller, it is true, for horticultural reasons that will be explained during the course in floriculture), but it should be understood that the flower is not invariably produced by the same bulb.
If we make a vertical cut on the bulb we will observe:
— the base and the dried vegetative unit that had produced the flower during the previous year,
— the base of the bulb that will produce the flower of the present year,
— the bud that will provide the vegetative unit of the flower during the next year.
The tulip is *a plant that flowers several times* with a vegetative unit that flowers only once, each older unit producing a new unit.
It is a perennial plant because of its bulb.
b) The bulb of the hyacinth (*Hyacinthus orientalis*) develops in a different manner.
The vegetative unit grows regularly and outwards from the centre, from a central bud. The external scales (the oldest) dry up and disappear.
The bulb extracted after the season is over is, therefore, not a new vegetative unit.
The hyacinth is a plant that flowers several times and has a pluriannual unit.

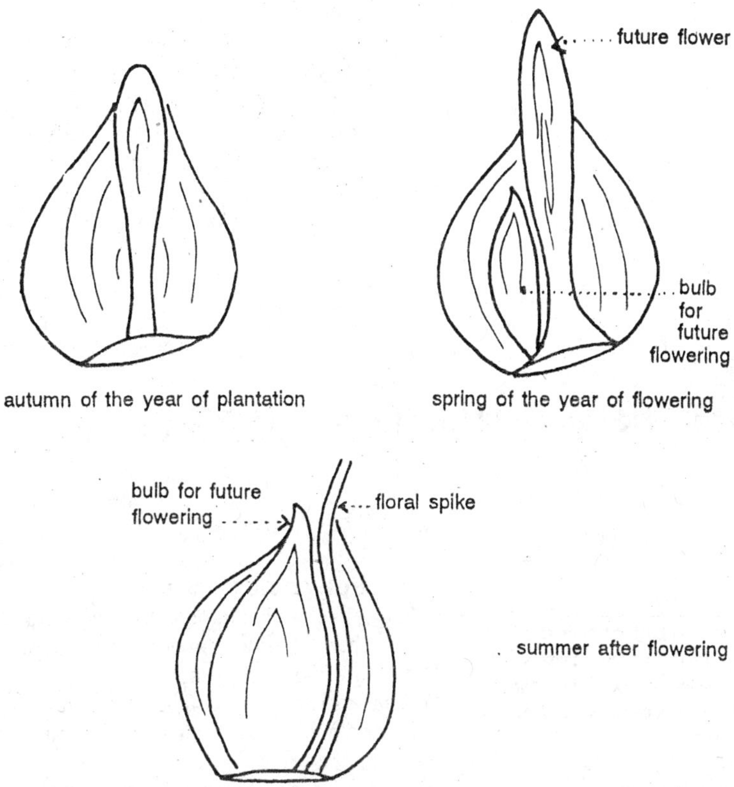

Fig. 2.72. Development of the tulip bulb

It is a perennial plant because of its bulb.

c) The lupin (*Lupinus polyphyllus*) produces a progressively larger vegetative unit every year with an increasing number of floral spikes.

Every year, the vegetative apparatus dies after preparing anlages of other vegetative organs for the coming years which, in turn, will flower and die.

The lupin is a plant that flowers several times with a vegetative unit that flowers only once, each older unit producing several new ones: this involves a multiplication.

It is a perennial plant because of its stock.

d) The pine (*Pinus sylvestris*) and the apple (*Malus communis*) have vegetative units that persist from year to year. They build up and branch.

They are *arborescent plants.*

The flowers on these plants, however, may appear on different branches:

— year-old branch: the vine (*Vitis vinifera*), buddleia (*Buddleia variabilis*);
— two-year-old branch: peach (*Prunus persica*);
— three-year-old branch: pear (*Pyrus communis*), in most cases. They may also be remontant when several flowerings occur on a year-old branch: as, for example, in the case of the remontant rose.

2
Morphology of Non-flowering Plants

2.1 The ferns

2.1.1 Polypodium aureum

We have observed ferns in the undergrowth of forests. Horticulturists grow several species of ferns, all of which are highly decorative.

Take the example of *Polypodium aureum* grown for the size and divisions in its leaves as well as for its bluish colour.

We will remove one from its pot and clear the soil that surrounds the roots.

We note:

a) Numerous fine roots.

b) What look like thick roots but, on examination, are found to be rhizomes (underground stems)[34] because they have both roots and leaves. Some parts of the rhizomes at the surface of the soil have many brown scales. These underground stems on the surface of the soil are branched.

c) Leaves that grow from rhizomes:

— Some, in the shape of a crosier, often covered with scales, are young.
— Others[35] of a large size have numerous divisions in the limb, with many veins. These are fronds (old leaves) that may be a metre long. They are glaucous on both surfaces.

d) There are never any flowers but, under the fronds, we observe brownish masses, the role of which will be studied when we examine reproduction in non-flowering plants.

34

Perennial because of its rhizomes, a fern may "grow again" after a partial or even complete cutting of its fronds. The stock produces new leaves.

35

Whatever the shape and length of the axes that carry the fronds, they are always petioles of leaves and not stems:

— Very long petiole of the *Pteris serrulata* with a blade that is often variegated.
— Fine brownish petiole of the *Adiantum cuneatum* (capillary) with a triangular blade.
— Very large frond (1.5 m) of *Nephrolepis exaltata* with a finely denticulated blade.

"Tropical" ferns are an exception, some of which are called arborescent, but what we see are actual trunks.

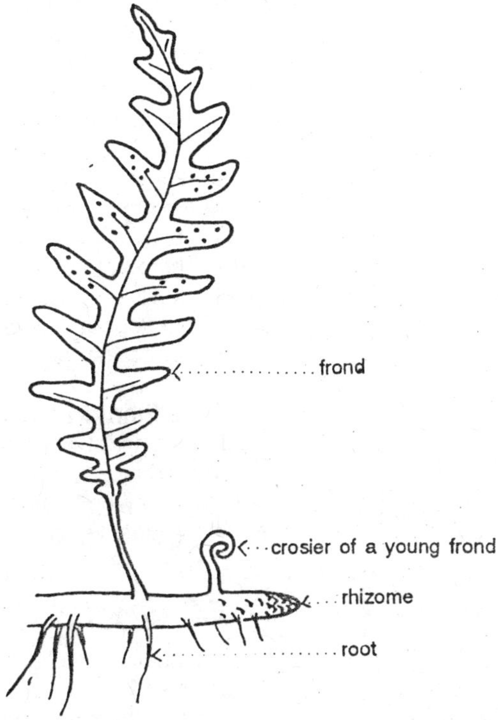

Fig. 2.73. *Polypodium aureum*

Morphological conclusions
Ferns have roots, stems and leaves.

2.1.2 Plants similar to ferns

a) Horsetails

These have the same organs as ferns but the underground stem (rhizome) also bears aerial stems that are slender, hollow and formed of detachable articles. The leaves are small, in whorls at the nodes on stems, sometimes with very slender branches, 2 to 3 cm long, that start from the whorl.

b) The selaginella

They have the same organs as ferns but the aerial stem is forked (dichotomous).[36]

36 The form of branching of stems in selaginella, the flat branches with their leaves and the stripes and variegations on these leaves lead horticulturists to cultivate this somewhat ornamental plant.

2.2 The mosses

2.2.1 *Sphagnum cymbiifolium*

All horticulturists are familiar with this moss, which

some use in their soil mixtures while cultivating members of Orchidaceae and Bromeliaceae.
Never more than 10 cm in size, it has a vegetative apparatus that is generally erect and comprises:
a) Young trailing filaments: protonemata.
b) The protonema has a self-supporting stem and
c) very small fine leaves of a greenish colour.
d) A protonema does not have roots but very fine hair-like organs known as rhizoids.
e) There are never any flowers but some short stems are fertile and bear reproductive organs.
Morphological conclusions. Mosses have stems and leaves but not roots.

2.2.2 Plants similar to mosses

Liverworts, of which the vegetative apparatus is usually trailing.
They may or may not have leaves; in the latter case the vegetative apparatus is reduced to a "thallus".

2.3 Fungi

The number of fungi is vast and diverse. Though some are visible to the naked eye and are grown as food plants, others are microscopic in size.
We will study two well-known examples:

2.3.1 The cultivated mushroom (*Psalliota campestris*)

Let us pay a visit to a mushroom grower and look at a "Paris mushroom" that is grown on a "mushroom bed", unless we can find a "wild" type in a meadow.
It is seen in several forms:
— small, young, in a stage of development,
— developed, older and . . . harvestable.

2.3.1.1 These whitish structures form:

— when they appear, an ovoid ball surrounded by an envelope;
— on growing, this envelope bursts and disappears, leaving at the base what is known as a volva;
— we then identify a stipe with a spherical shape surrounded by a velum or veil;
— as it grows this velum is torn and its fragments remain on the stipe to form a ring;
— the cap then appears in its final shape.
All this development takes place in two or three days.

The white cap, on its lower surface, has a row of vertical lamellae that are, initially, of a light colour and become darker as they age.
This "edible" production of the cultivated mushroom is its reproductive organ.

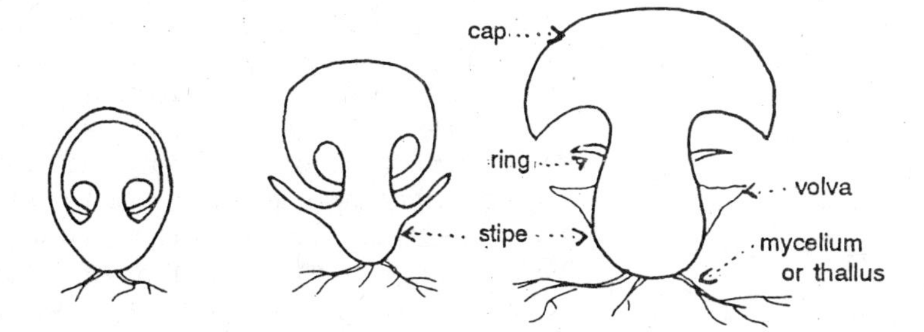

Fig. 2.74. "Aerial" development of the cultivated mushroom

2.3.1.2 Let us uproot a stipe and its cap

We observe a collection of partitioned filaments: this is the thallus of the mushroom, also known as the mycelium.
In the cultivated mushroom it is also called the "spawn".
This "inedible" portion of the cultivated mushroom is its vegetative organ.
Note: The chanterelle (skirret), the parasol mushroom, and St George's agaric have the same organs as the cultivated mushroom, but they look different.
On the other hand, the *Boletus* or cepe does not have lamellae but tubes on the underside of its cap.
Important: We consider it necessary to draw attention to the *morphological appearance* of capped mushrooms, *which alone enables a systematic determination of edible mushrooms* and thus avoidance of any dangerous or even mortal error when sorting harvested products.
We do not need to bear any other rule in mind. Among others, we should be aware of a well-known mushroom called the Deadly Amanita.

2.3.2 Scab on the apple tree (*Venturia inacqualis*)

We note the "effects" of this cryptogamic infection on apple trees in the form of:

— brown stains on the leaves,

— cracks on the branches, fruits, etc.

But we must use a magnifying glass or even a microscope to study the morphology of this fungus.
In fact, the "mycelium" is not visible. It develops under the epidermis of the infected organs.
Under the microscope we observe a downy tangle of filaments: the thallus.
Using a magnifying glass we observe "fructifications" resulting in the germination of "seeds": the spores, which contaminate other young organs.
This parasite can survive easily during the winter in the form of "reserve" reproductive organs, a type of eggs, found on dead leaves and on the trunks of trees ... that ensure contamination in the spring, in other words, the first germination from which the new mycelium will emerge.
The mycelium, here again, develops internally.
Only the fructifications are external as in the case of the cultivated mushroom.[37]

37
The arboriculturist who wishes to protect his orchard must prevent the penetration of this parasite because it develops internally.
Treatment can only be preventive.
Curative treatments are called for only as a last resort for a "disastrous" case except in a few situations when the development of the fungus is external.

Morphological conclusions
Fungi have no roots, stems or leaves.
Their vegetative apparatus is merely a thallus, most often in the form of filaments, that may be microscopic or even several centimetres in size.

2.4 The algae

2.4.1 Brown seaweed

We are familiar with brown seaweed or kelp, which is observed on Atlantic beaches, especially in Brittany.
Members of Fucaceae and Laminariaceae, these algae are seen in the form of brownish strips of a fairly large size. Several decimetres long, branched and silky to the touch, they are known as brown algae.
They are attached to rocks by crampons that look like roots but from the botanical point of view are not roots.
The floating parts have floats or vesicles filled with air.
The ends of certain branches have reproductive organs.

2.4.2 Microscopic algae

Among others, in the brown algae we find diatoms covered with a protective shell that, together with living microscopic animals, form plankton.
Morphological conclusions: Algae have no roots, stems or leaves and, like fungi, have a vegetative apparatus that is reduced to a thallus, which may be microscopic or large.

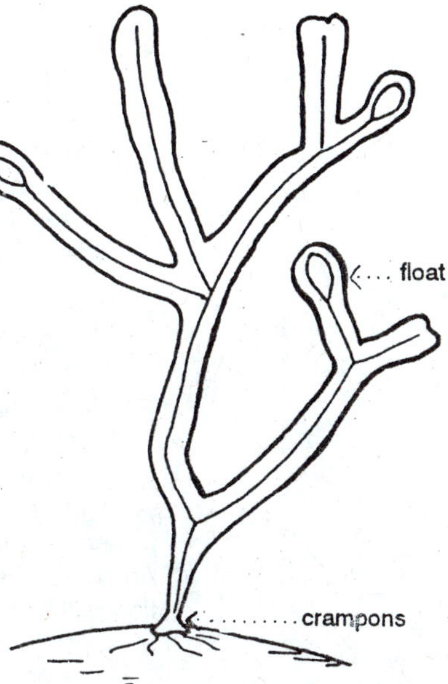

Fig. 2.75. The fucus

2.5 Lichens

On walls that are usually damp, we may observe small, scaly green blades that are somewhat lobed.
These are wall parmelia, which are also found on the bark of trees.
They are attached by rhizoids that are not roots.
At certain points on their surface there are reproductive organs that, to some extent, resemble those of fungi.
Morphological conclusions: Lichens have no roots, stems or leaves, and their vegetative apparatus is merely a thallus.

3 Morphology and Systematics

In this chapter we have observed and described flowering plants, the phanerogams, and non-flowering plants, the cryptogams.

We can morphologically identify:

■ *The phanerogams* (flowering or seed plants)
— without an ovary, with exposed ovules: gymnosperms,
— with an ovary and, therefore, covered ovules: angiosperms, with seeds containing one cotyledon: monocotyledons and with seeds containing two cotyledons: dicotyledons.

■ *The cryptogams* (non-flowering plants)
— Plants that have roots, stems and leaves: pteridophytes (ferns, horsetails, selaginella).
— Plants that have stems and leaves but no roots: bryophytes (mosses, liverworts).
— Plants that do not have roots, stems or leaves: thallophytes (fungi, algae, lichens, bacteria), in which the vegetative apparatus is a thallus.

An anatomical study, especially in respect of reproduction, will help us to complete this classification scheme of plants (*see Chapter 8, "Systematics"*).

Chapter 3

Cells and Tissues

1 Microscopic Observations

Our study of morphology is based, mainly, on observations of several plants.
To study plant cells and tissues, we need to observe "microscopic" organs.

1.1 Equipment

For this purpose we need:
— a magnifying glass,
— a microscope with one or several object lenses.
The microscope, an expensive item, must be properly maintained, especially its optical parts:
— prevent it from being scratched,
— never dismantle it, and operate the fine-adjustment screw carefully to avoid damaging the observation slide.

1.2 Preparations

a) The sections should be as thin as possible. They may be made:
— either with a razor blade: the organ to be sectioned can be held in "elder pith";
— or by removing a strip of the organ if the cells are large.
b) The sections should then be treated, depending on the case, with various reagents and stains, etc. They should then be successively:
— immersed for 10 minutes in Javel water to destroy the contents of the cells, leaving only their membranes;
— washed in acetic acid that will neutralise the Javel water;
— rinsed in distilled water;

— stained with carmine that will turn cellulose pink or with iodine green that will render wood, suberin and cutin green.

These operations are to be carried out in a watch glass with the help of small forceps.

c) The sections will be placed between:

— a glass slide and

— a smaller and much thinner coverslip, with the object usually kept in a drop of water.

1.3 Observations

a) Regulate the lighting with the help of a mirror (usually the concave side). Concentrate the light to obtain a good view.

b) Adjustment:

— first, with the rack to obtain a small image;

— next, with the fine adjustment screw for "clarity"

— after this use the high-power lens to study the details.

2
The Cell

2.1 Description of a cell

2.1.1 Let us remove a strip of the skin that rises when we break the internal scales of a hyacinth (*Hyacinthus orientalis*) bulb.

When examined at low power under a microscope, this epidermis is seen to be formed by a group of elongated polyhedral elements that adhere to each other.

Inside, we note a spherical organelle: *the nucleus.*

This nucleus lies in the mass of the cell: *the protoplasm.*

The protoplasm contains many granulations as well as a fairly large number of cavities of various sizes: *the vacuoles.*

Externally, the protoplasm is surrounded by a plasma membrane that in turn is covered by a second, thicker membrane called the cell wall.

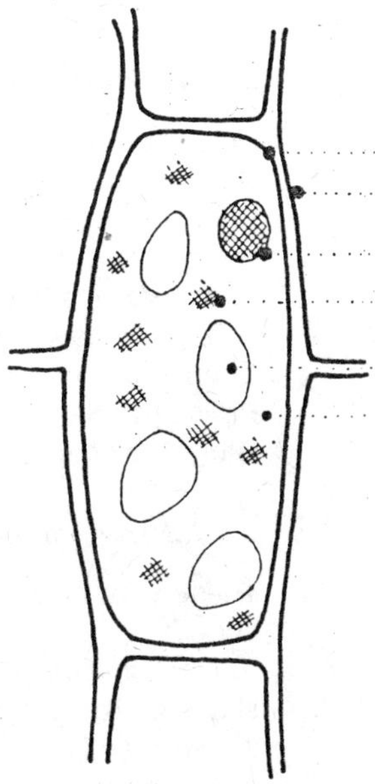

Fig. 3.1. The cell

2.1.2 If we, in the same manner, examine leaves of iris (*Iris hybrida*), in which epidermal cells are as easily detached as those in the hyacinth bulb, we observe:

— A nucleus.
— Protoplasm with many vacuoles and granulations, but in this case we also find several green structures containing chlorophyll: these are chloroplasts.
— Membranes.

2.1.3 Finally, in a small portion of tomato pulp (*Solanum lycopersicum*) we find the same parts as we do in the cell of a bulb; of course, there are no chloroplasts as in the iris but there are reddish grains: these are chromoplasts containing carotene.

2.2 Different parts of a cell

Based on our previous observations we may conclude that cells are of very varied forms and comprise:

— a nucleus,
— a protoplasm containing vacuoles of different sizes and various granulations, and
— two membranes.

2.2.1 Protoplasm

2.2.1.1 Composition

Protoplasm is composed of a transparent albuminoid substance that is fairly consistent, resembling the white of an egg with a somewhat similar chemical composition, in other words: 75% water, glucosides, lipids, proteins, organic acids and mineral elements.
(The definitions of these compounds are given in Chapter 5, §1.6.2).
Protoplasm is sensitive to heat and undergoes circulatory movements.

2.2.1.2 The granulations

These vary in both shape and appearance. We observe:
a) Chondriosomes: especially *mitochondria* in the shape of filaments, rods or grains.
These granules are made of lipids and proteins and are collectively called the *chondriome*.
b) Plastids that accumulate various substances.
When colourless they are known as *leucoplasts*.
When they fix starch, they are called *amyloplasts*.
When they fix oils, they are *oleoplasts* or *elaioplasts*.
When yellow or red, they are *chromoplasts*.
If green, they are *chloroplasts* or "chlorophyll grains".
Leuco-, amylo- and oleoplasts play an important role in the storage of nutritive reserves.
Chromoplasts determine the colour of organs in plants, other than the green colour provided by the chlorophyll present in chloroplasts.
All these plastids are collectively called the *plastidome*.

2.2.1.3 Vacuoles

These are spaces filled with liquid or cellular sap in which, among other things, we find water and mineral salts drawn from the soil by the plant for its nutrition.[1]
It may be noted that:

— When a cell is young it is filled completely with protoplasm.
— When it is old, the protoplasm becomes hollow with progressively larger cavities or vacuoles; at the same time the nucleus is forced towards the cell wall along with the remaining protoplasm.
— When the entire protoplasm disappears, the cell is dead.

1 If a plant does not have enough water the turgidity of the vacuoles is reduced and the plant wilts.

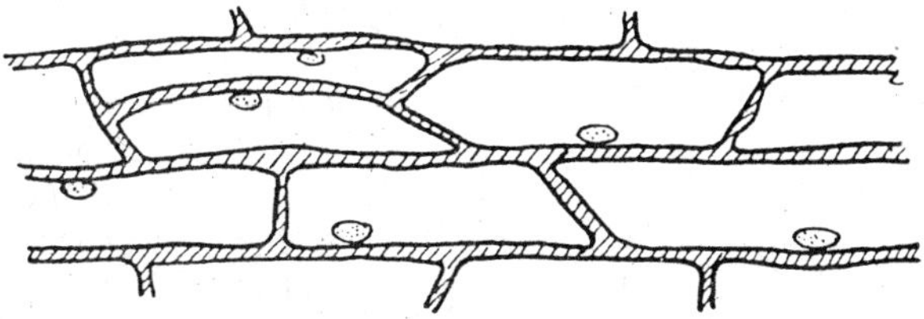

Fig. 3.2. The cells have grown old: the nucleus is forced against the cell wall

2.2.2 The nucleus

Generally spherical, it is composed of an albuminoid substance similar to that of the protoplasm but with a thicker consistency, called *nuclear sap* and surrounded by a thin membrane: the *nuclear membrane.*

In the nuclear sap there is a network of coiled filaments known as the *chromatin reticulum.*

During the process of cell division this chromatin reticulum breaks up into a certain number of chromosomes.

Chromosomes carry the hereditary characteristics of an individual and their number is always fixed for an organism.

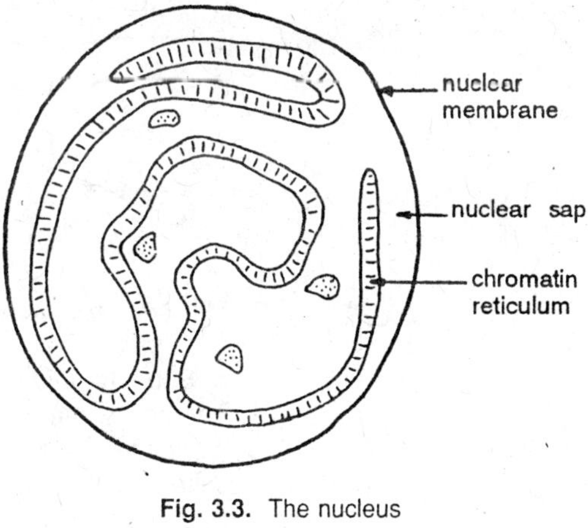

Fig. 3.3. The nucleus

2.2.3 The membranes

We have seen that there are two membranes in a cell.

a) Internally, the protoplasm is, to some extent, held in a thin skin as a result of its physical modification: this is the *plasma membrane.*

b) Externally, a thicker membrane covers the plasma membrane. It is composed of cellulose; cellulose is a hydrocarbon somewhat similar to starch, permeable by water and dissolved substances.
This property enables the plant to absorb raw sap (*see Chapter 5, "Physiology"*).
It is soluble in Schweitzer liquor (ammonia + copper), an experiment that is easy to carry out with cotton wool.
Carmine stains it pink.
This membrane is called the *cellulose membrane* or *cell wall.*

2.2.3.2 Modifications in the cellulose membrane or cell wall

The cellulose in the external membrane may be transformed into other substances or be encrusted with products elaborated by the protoplasm.
It may:
a) Be cutinised: *this is cutinisation.*
The organs exposed to the air acquire a greater hardness and a certain impermeability in order to resist desiccation of the cell during hot weather and its destruction when temperatures are very low.
These organs are covered with a layer of *cutin*, a fatty substance: this is seen in the waxy cuticle of holly (*Ilex aquifolium*).

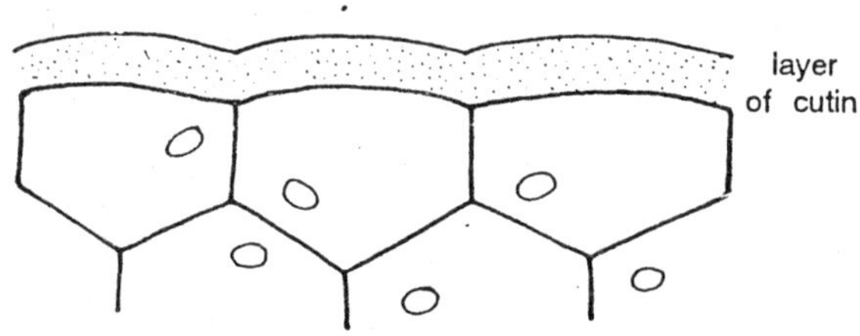

Fig. 3.4. Cutinisation of the cellulose membrane or cell wall

b) Suberify: *this is suberification*
The cell is surrounded with cork "suberin"; cork forms a fairly thick sleeve composed of various layers of flat cells arranged in radial rows.
Suberin, like cutin, forms a protective coating on organs that have reached their "adult" state.
This applies to the oldest parts of roots and old stems.
In some cases cork formation is considerable: for example, the cork oak (*Quercus suber*).
c) Lignify: *this is lignification*
The cell wall is encrusted with lignin, a brittle brown

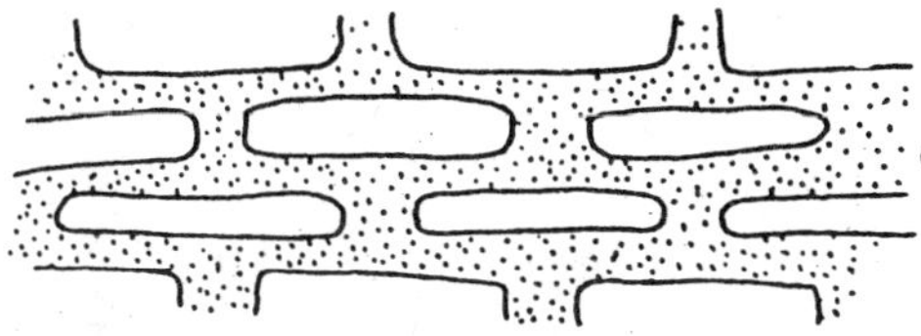

Fig. 3.5. Suberification of the cellulose membrane or cell wall

material, rich in carbon, that increases the resistance of the cellulose walls.
This is seen in the case of a herbaceous stem which "matures" and becomes ligneous.
Lignin is stained green by iodine green.[2]
d) Gelify: *this is gelification*
Cellulose is transformed into a type of jelly formed of gums, as in cherry tree (*Prunus cerasus*) and peach tree (*Prunus persica*), or of mucilages, as in the fruit of the tomato (*Solanum lycopersicum*), that may even dissolve completely (*fruits at maturity*).
e) Mineralise: *this is mineralisation*
The wall is impregnated with silica—as in graminaceous plants—or with limestone—as in algae, etc.

2
For espaliating and shaping fruit trees in a nursery, for example, we should not wait till the branches have become highly lignified: bendings will then only be more difficult to carry out.

2.3 Cell division

Every plant develops from a single cell: the egg cell (*see Chapter 6, §§ 1.6 and 1.7*).
A cell grows and divides into two daughter cells that may exist:
either separately: unicellular plants (bacteria ...) or in association, to form a group of cells: multicellular plants.
Cell division is a physiological act of a plant, highly complex and very active in the younger parts of organs known as initial cells.
We can identify:

2.3.1 Amitosis

Division takes place by a single constriction: e.g., bacteria, yeasts.

2.3.2 Mitosis

Division takes place only after division of the nucleus, an operation that is known as karyokinesis.
This is the most common process.
Karyokinesis comprises a certain number of phases.

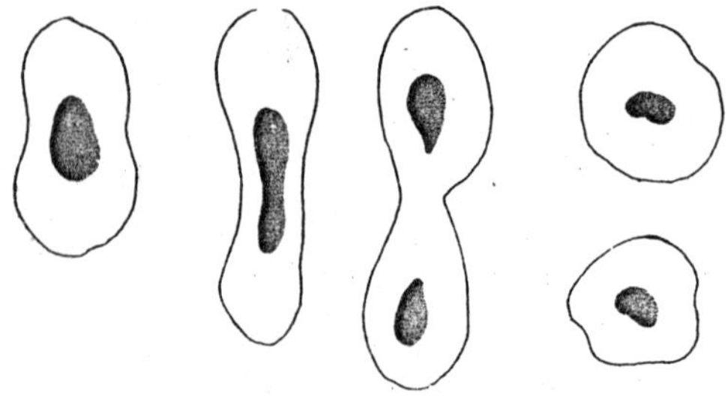

Fig. 3.6. Amitosis of yeasts

1) The *directive sphere*, which is always near the nucleus, divides into two spheres that distance themselves from each other and move to the two ends of the same diameter of the nucleus.
The nuclear membrane disappears.
2) The chromatin reticulum of the nucleus is segmented in a certain number of fragments—the *chromosomes* ($2n$)—a number that is fixed for each species. At the same time, filaments appear that are types of granular trails extending from one pole of the cell to the other, forming a type of *spindle*.
3) The *chromosomes* in the shape of a U or V *divide into two*, lengthwise.
The twin chromosomes then move along the filaments of the spindle and *proceed* in the opposite direction *towards one of the directive spheres*.
4) Each group of $2n$ chromosomes joins up, end to end, and *again forms a chromatin reticulum*.
5) A *nuclear membrane* surrounds each chromatin reticulum and, in this way, two new nuclei are formed.
6) Once karyokinesis is complete, there is a constriction at the centre of the cell and two cells are then formed from the mother cell.
We now have two daughter cells, each with a protoplasm, a membrane, a nucleus and a directive sphere. The nucleus has $2n$ chromosomes: the plant is diploid. The cells grow and will, in turn, divide.
The entire process is usually over in 1 to 2 hours and is known as *equational mitosis*.
In certain particular cases, however, the chromosomes do not divide into two. Each daughter cell has only n chromosomes.
This happens in the division of sex cells that produce male and female elements (gametes).
This particular process is called *reductional mitosis* or *meiosis*.

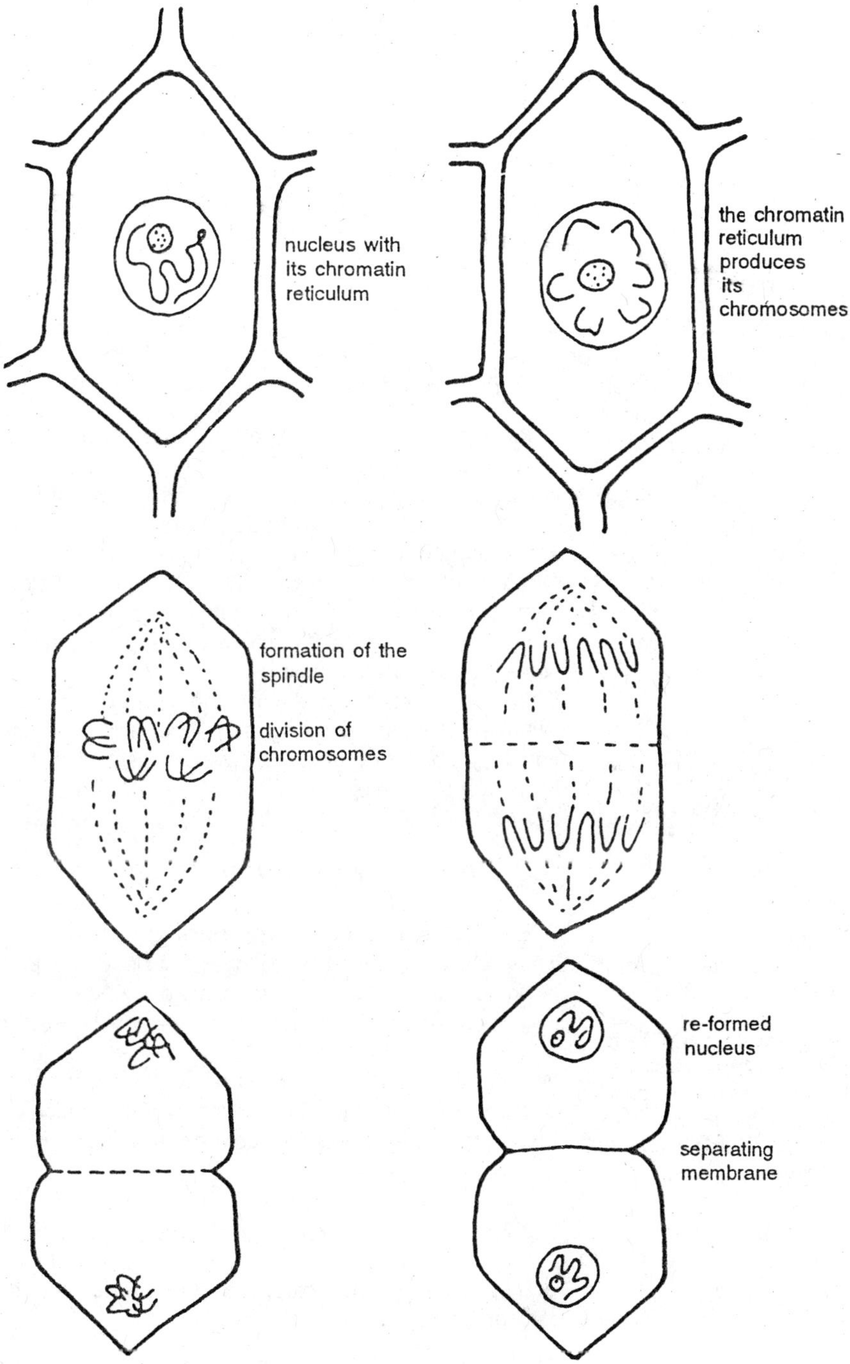

Fig. 3.7. Mitosis or karyokinesis

It may be noted that on formation of an egg resulting from the fusion of a male gamete and a female gamete, this egg will again have $2n$ chromosomes.

3 Tissues

3.1 Definition

In lower plants that have reached an adult age, the entire plant body is composed of cells. Each cell resembles its neighbour and functions on its own without adversely affecting the plant as a whole: for example, some algae or certain fungi.
On the other hand, in higher plants the cells develop in many different ways and indicate varied characteristics. We say that they differentiate; they organise themselves for a specific function.
They form a tissue: wood cells, cork cells, etc.
When several tissues combine to perform a vital function *they constitute an organ*: various tissues of the roots, the stem, etc.

3.2 Different types of tissues

3.2.1 Young tissues or meristems

These are very young tissues formed of closely packed living cells with no intercellular spaces.
These cells multiply actively to ensure the growth of organs.
We find them mainly in seeds as well as in the tips of roots and branches, for ensuring an elongation of these organs, as well as inside them to enable an increase in thickness.[3]

3 The horticulturist can even ensure multiplication of some plants from cuttings of meristems, for example, carnation (*Dianthus caryophyllus*), chrysanthemum (*Chrysanthemum indicum*).

3.2.2 Mature or permanent tissues

3.2.2.1 Tissues that carry out vital functions or parenchyma

These are formed of living cells inside which all vital functions of a plant are carried out.

We observe:
a) Chlorenchyma or *assimilative tissue* formed of cells rich in chloroplasts, especially in the pallisade tissue towards the upper surface of leaves.

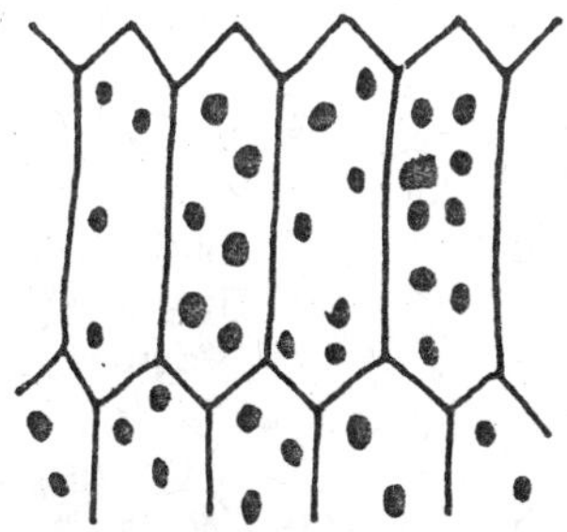

Fig. 3.8. Chlorenchyma in the pallisade parenchyma of leaves

b) Colourless tissue that we find in underground organs (roots, bulbs, tubers ...) often full of reserve materials.
c) Secretory tissue in which the cells secrete products considered to be the waste material of a plant.
These secretions are of various types:
— resins of conifers;
— essential oils of sage (*Salvia splendens*), basil (*Ocimum basilicum*), flowers (petals) of the rose (*Rosa*);
— latex of the poppy (*Papaver somniferum*), euphorbia (*Euphorbia pulcherrima* or Poinsettia), lettuce (*Lactuca sativa*), rubber (*Hevea brasiliensis*);
— alkaloids of tobacco (*Nicotiana tabacum*), coffee (*Coffea arabica*);
— sugar, nectar of flowers.
This function is carried out:
— either by secretory cells
— or by secretory canals (hairs ending in a gland).

3.2.2.2 Purely mechanical tissues

These play no part in the vital functions of a plant and have only a mechanical role.
a) Protective tissues
■ *Epidermis*
A primary tissue formed by the outermost row of cells surrounding the organs.
The external wall of this row of cells is covered with cutin and forms a wax cuticle. This is known as "bloom"

If the cuticle is thick the organ is rigid—leaf of the box tree (*Buxus sempervirens*)—and this helps in resistance to cold and dry weather.

On the surface of the epidermis we often find:

— projections: papillae on flower petals;

— hairs: hairlike secretory structures on nettles (*Urtica dioica*);

— stomata (*see Chapter 4, § 1.3*).

■ *Cork*

This is a secondary tissue composed of flat dead cells. They make up for the disappearance of cells of the epidermis that are not "elastic" and "tear" when a root or a stem increases in size.

All tissues on the outside of the root or stem die.

Cork is also formed to ensure proper healing.[4]

4 The formation of cork when there is accidental damage (breakage) or deliberate cutting (of a large branch) is not rapid. It is, therefore, desirable to protect the cut portion with a mastic.

b) Sap-conducting tissues

■ *Xylem vessels: vascular tissue*

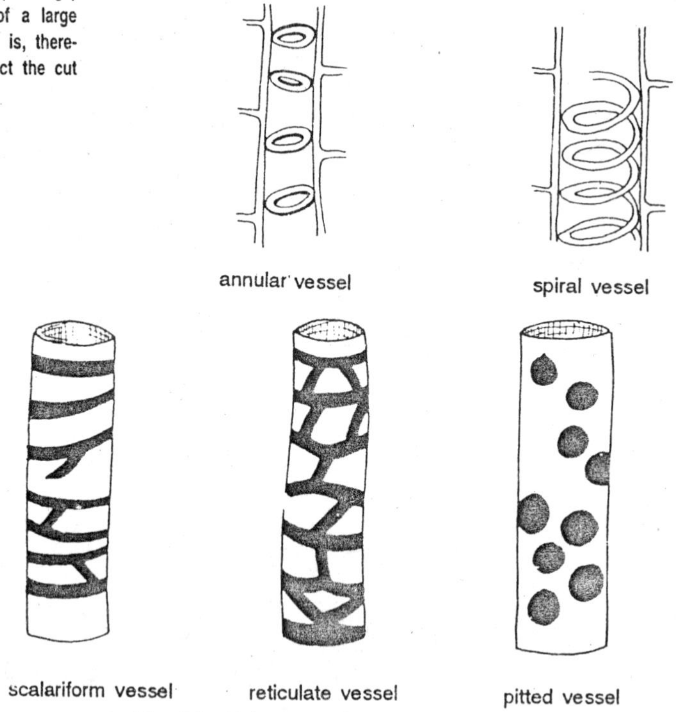

Fig. 3.9. Xylem vessels

They are formed from xylem elements placed end to end.

These vessels conduct raw sap (water + mineral salts) from the roots to the leaves.

The walls of these vessels are encrusted with lignin in different ways that give them varied appearances:

— initially, annulated vessels;

— then, spiral;

— of a primary formation that is observed mainly in young tissues because they can increase in size.

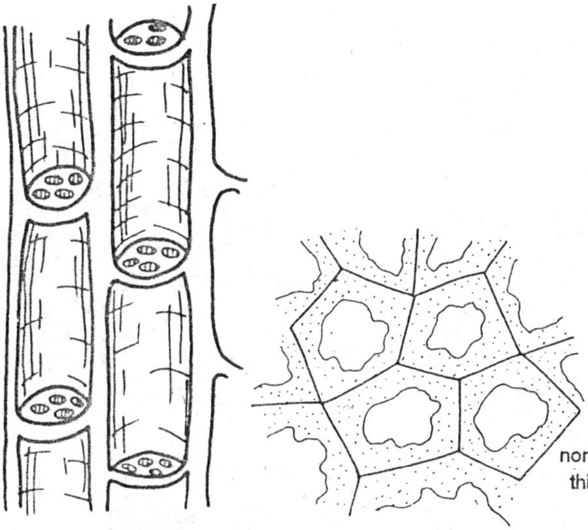

Fig. 3.10. A sieve tube for elaborated sap
The collenchyma

Subsequently, the vessels are scalariform, reticulate and, finally, pitted.

—of a secondary formation as is observed in adult tissues that do not grow larger.

In the beginning, these vessels still have transversal partitions of cells. They are imperfect vessels.

Subsequently, these partitions dissolve and disappear, and the vessels become perfect vessels.

Note on wood elements of conifers

These plants do not have different types of vessels. They have vascular tubes formed from imperfect elements known as tracheids with lateral perforations.

■ *Sieve tubes: phloem tissue*

These are in the form of rows of living cells but without nuclei and are present in all parts of the plant.

They conduct elaborated sap from the leaves to all areas where they are required.

At different places, there are transverse partitions riddled with holes, called *sieve plates.*

These sieve plates act as a strainer and help viscous elaborated sap to circulate by draining (experiment on the vine).
During the autumn they are blocked by a callus of cellulose that usually disappears in the spring.
c) Supportive tissues
These tissues give the plant a certain rigidity. Composed of living cells with thick but not lignified cellulose walls, they form the supple but resistant collenchyma. (Reinforcement of the angles of herbaceous stems in Labiatae).
They may, sometimes, form fibres that, when long, provide textile fibre (fibre flax).

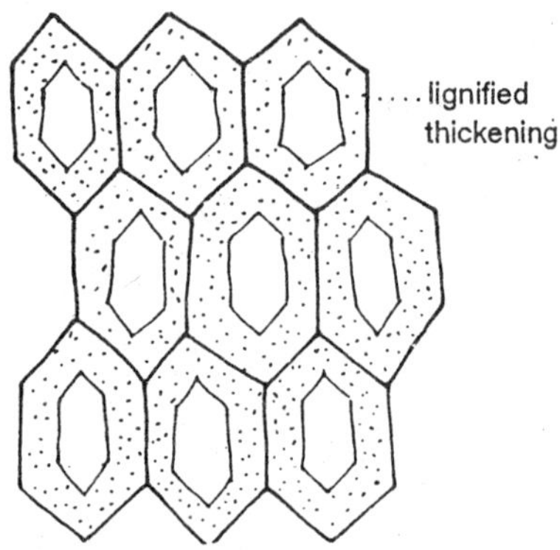

Fig. 3.11. Sclerenchyma

Formed of dead cells with lignified walls and so thick that the cellular cavity has completely disappeared, they constitute hard and brittle *sclerenchyma*.
Peach stone (*Persica vulgaris*), cherry stone (*Cerasus avium*).
They are sometimes observed in the form of cells ("stony" winter pear or *Pyrus communis*) or fibres (tree wood).

Chapter 4

Anatomy of the Vegetative Parts of Plants

1

Anatomy of the Vegetative Parts of Phanerogams

Since an anatomical study of these plants is based primarily on observation, we need to examine a certain number of "microscopic sections" of the organs we wish to investigate.
We must, therefore, observe the internal organisation of tissues and be able to differentiate them as we have described in Chapter 3.
The sections will have to be prepared as indicated in Chapter 3, §1.2.

1.1 Anatomy of the root

We are aware that there are two successive structures in a root: a primary and a secondary structure.

1.1.1 Primary structure

Let us observe a transverse section of a young root of the apple tree (*Malus communis*) in the region of absorbent hairs.
Moving inwards from the outside we can identify:
a) A layer of regular cells, some of which are extended in the form of absorbent hairs. In these, the nucleus is at the end of the hair-like structure: *this is the piliferous layer.*
b) A region of closely packed cells: *this is the cortical zone.*
c) A layer of cubic cells that are somewhat suberised: *this is the endodermis.*
d) A layer of regular cells alternating with those of the endodermis: *this is the pericycle.*
e) An alternate arrangement of sap-conducting strands:

— some roundish and grouped in oval masses, of a pink colour,
— others longer and grouped in a triangle with the apex facing outwards, of a green colour.

The former constitute phloem.
The latter constitute xylem.
They arè arranged alternately.

f) A mass of filler cells in the centre: this is the pith, which ramifies between the phloem and xylem strands form *the medullary rays.*

—The piliferous layer is the outermost layer —The cortical zone is the thickest —The endodermis is the innermost	which form parts of the bark
—The pericycle is the outermost layer —The phloem and xylem strands —The pith	which form parts of the central cylinder

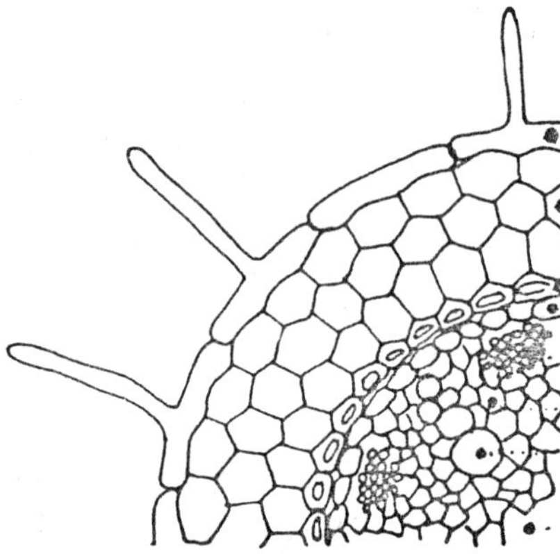

Fig. 4.1. Primary structure of the root

1.1.2 Secondary structure: Increase in thickness

We now examine a transverse section of a root from an apple tree taken, this time, from an older part which had thickened.

1.1.2.1 How does a root increase in thickness?

By the action of two generative layers.

A generative layer is a layer of cells operating by successive divisions in which younger cells displace older ones.

Moreover, it operates in both directions: internally and externally.

1.1.2.2 Internal vascular generative layer or cambium in the central cylinder

It is situated:
— opposite to and in the inner side of the phloem,
— opposite to and outside the xylem.
It is, therefore, sinuous.
It forms xylem on its internal side and phloem on the external side.
In the spring a large amount of sap circulates; the vessels are wide and of a light colour.
In the autumn the vessels are narrower and darker.

1.1.2.3 External cork cambium generative layer or phellogen found in the bark

It is located outside the cortical layer.
It is circular.
It produces cork on the external side and phelloderm (parenchyma) on the internal side.

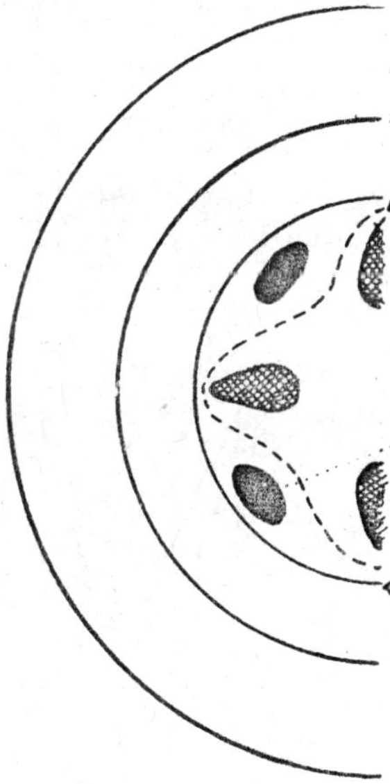

Fig. 4.2. Position of cambium layers in the root

1.1.2.4 What do we identify in a section of a secondary root?

From the outside, moving inwards:
a) In the bark:

cork that has replaced the piliferous layer,
the external generative layer or cork cambium,
the phelloderm,
the primary cortical zone,
the endodermis.
b) In the central cylinder:
the pericycle,
the primary phloem,
the secondary phloem,
the internal generative layer or vascular cambium,
the secondary xylem,
the primary xylem,
the pith.
It may be noted that, in most monocotyledons, secondary formations do not appear.

1.1.3 Increase in length

Roots increase in length to various extents and the differences may be considerable:

— A few centimetres in the alpine forget-me-not (*Myosotis alpestris*).

— Several metres in the apple tree (*Malus communis*) and the Scotch pine (*Pinus sylvestris*).

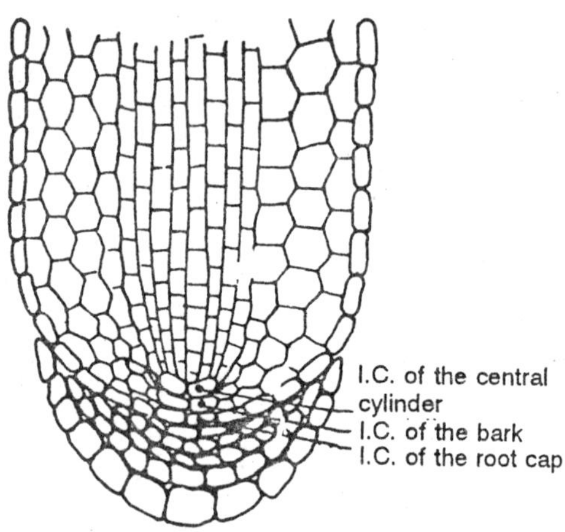

Fig. 4.3. Initial cells that enable elongation of a root

This growth is due to the action of three initial cells. Let us make a longitudinal section of the end of a root. We identify three starting points for this increase in length:

— a cell that forms the central cylinder,

— a cell that forms the bark,

— a cell that forms the piliferous layer and the root cap.

The cell that forms the root cap is regenerated as and when its tissues are worn down due to penetration in the soil.

We understand why this lengthening occurs in the sub-terminal region of the root.

1.1.4 Branching of the root

The main root divides into secondary roots and these may also be branched.

The rootlets are not arranged in a haphazard manner but follow the generative lines opposite the xylem or phloem elements or, simultaneously, both of them—except, obviously, in the case of adventitious roots.

The origin of rootlets is endogenous, which means they originate from the central cylinder and pierce through the bark before they emerge from the root.[1]

1 Injuries on roots are cicatrised by an emission of cork and this protection is ensured even more rapidly if the injury is clean and open.

1.2 Anatomy of the stem

1.2.1 Primary structure

Let us examine a transverse section of a young stem from an apple tree.

Moving inwards from the outside we observe:

a) A layer of cells in which the external protective wall is cutinised (cuticle): *this is the epidermis*. It has stomata, but there are not many of them.

b) A region of cells arranged in a somewhat regular manner but with intercellular spaces: *this is the cortical zone.*

This zone contains:

— *chloroplasts* (whence its role in chlorophyll assimilation);

— *collenchyma* distributed in the external portion of the cortical zone (reinforcement of the angles of stems in Labiatae).

c) A layer of cells as in the root: *the endodermis.*

d) A second layer of cells—*the pericycle*—alternating with those of the endodermis.

It may be noted that endodermis-pericycle differentiation is not as clear as in the root and, further, these two layers are not found in monocotyledons.

e) A difference from the root: the vessels of xylem and phloem are superposed and conjoint. They form the triangular *vascular bundles* with the apex turned towards

the centre, where the phloem is towards the outside (pink) and the xylem towards the inside (green).
f) A filler tissue, *the pith*, formed between the vascular bundles: *the medullary rays.*[2]

—the epidermis is the outermost layer —the cortical zone the thickest —the endodermis the innermost	which form parts of the bark
—the pericycle is the outermost layer —the vascular bundles —the pith	which form parts of the central cylinder

2 A tree trunk is easy to split open if we follow the radial axes: these are the medullary rays, which are less resistant than the vascular bundles.

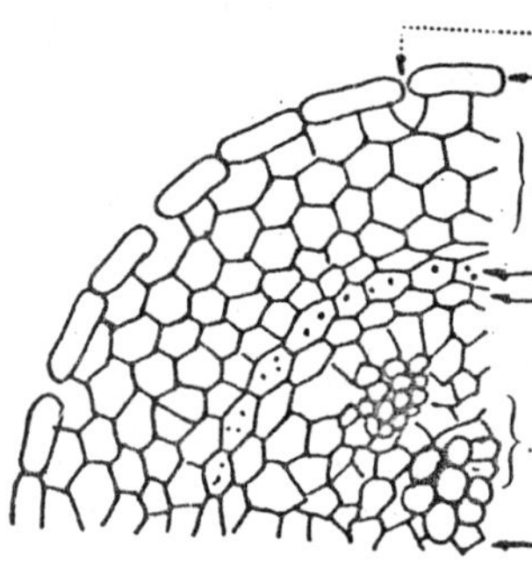

Fig. 4.4. Primary structure of a stem

To conclude: in the primary structure we find the following:
In the stem:
the epidermis replaces the piliferous layer seen in the root, the cortical zone has intercellular spaces that are not found in a root, the endodermis and the pericycle are not distinct.
Contrary to what is observed in a root, the xylem and phloem elements are superposed (they are arranged alternately in a root).
Since the main difference between a stem and a root lies in the arrangement of the vascular bundles, how is their "junction" achieved?
The structural change occurs in *the collar.*
Just below the collar, in the root, xylem strands are observed opposite phloem strands and, simultaneously, the number of normal xylem strands decreases.
Just above the collar, in the stem, the normal xylem strands of the root disappear and, at the same time,

there is an increase in those that appear opposite the phloem strands.

1.2.2 Secondary structure: increase in thickness

Let us examine a transverse section of an old stem taken from an apple tree.

1.2.2.1 The action of generative layers in the stem is the same as that of generative layers in the root and, in the stem, begins within a few days and, sometimes, a few weeks.

1.2.2.2 Internal vascular generative row or vascular cambium seen in the central cylinder

a) This is situated between the xylem and phloem elements.
It is, therefore, circular and not sinuous as in the root.
It forms secondary phloem externally and secondary xylem or wood internally.[3]
b) It is not very active in annual plants.
On the other hand, it is active in arborescent plants, especially in respect of wood.[4]
c) Its activity varies from one species to another and the rate of formation of wood varies.[5]
d) As in the root, the rate of growth is seasonal.
There is a considerable circulation of sap in the spring: the zone of secondary xylem or phloem is light coloured and not very compact.
In summer the flow of sap is reduced: the zone that is formed more slowly has thicker walls in the vessels. The zone is harder because the *support* function is more important than the *transport* function.[6]
e) The older portions of wood are in the centre. The younger portions are towards the outside.[7]

1.2.2.3 External generative layer or cork cambium

This is located outside the cortical layer.
It is circular.
It produces:
a) Cork towards the exterior that tears and forms lenticels (a type of stomata in cork).
The oldest ones fall regularly:
either in thin scales: cherry (*Prunus avium*), birch (*Betula alba*);
or in pieces: oak (*Quercus pedunculata*);
or in rough hair-like structures: vine (*Vitis vinifera*).

3
When we "strip" a stem we remove a ribbon of tissue that extends from the epidermis to the cambium. Only wood and pith, therefore, remain on the stem.
Under such conditions raw sap can continue to circulate from the roots towards the leaves but elaborated sap cannot circulate below the stripped portion.
This operation is carried out to assist the development of organs located above the decorticated area.
On the other hand, a deep incision also cuts the wood vessels and is carried out to assist in the growth of organs located below the incision.

4
Most of the trunk is composed of wood.

5
When wood formation is rapid the vessels are thick: the wood is soft (poplar: *Populus nigra*).
If wood formation is slow the vessels are thinner and closely packed: the wood is hard (oak: *Quercus pedunculata*).

6
We can calculate the age of a trunk by counting the annual growth rings, each containing a portion of spring wood and a portion of summer wood.

7
The centre of a trunk is hard: this is the heart wood.
Towards the outside the wood is softer: it is the alburnum (or sapwood), more porous and lighter in colour.
As it grows older sapwood is transformed to heartwood as a result of a chemical change.

b) Secondary parenchyma towards the inside: the phelloderm.

1.2.2.4 What do we observe in our section of a secondary stem?

From the outside moving inwards:
a) In the bark:
— a dead corky epidermis that may or may not fall away,
— a supportive tissue,
— primary bark,
— the internal generative layer or cork cambium,
— the secondary cortical parenchyma,
— the remains of the primary cortical parenchyma.
The endoderm and the pericycle are not clearly demarcated.
b) In the central cylinder
— the primary phloem
— the secondary phloem,
— the internal generative layer or vascular cambium,
— the secondary xylem,
— the primary xylem,
— the pith.

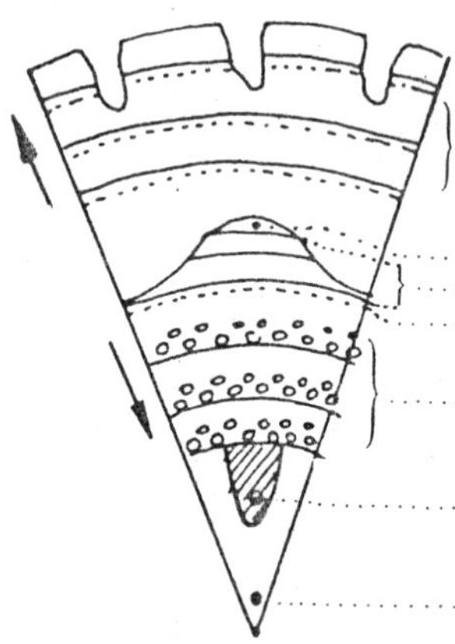

Fig. 4.5. Secondary structure of a stem

It may be noted that in most monocotyledons there is no secondary growth.
There is no cambium between the xylem and phloem. The thickening of vascular bundles is replaced by an increase in their number.
— In gymnosperms the wood is homogeneous because there is only a single growth, the spring

growth, and therefore a single appearance of tracheids.

— Every injury on the stems of plants (deliberate by pruning, or involuntary by accident) should be cicatrised by an emission of protective cork against air, humidity and diseases.[8]

8 To assist in formation of a cicatrice, cuts should be clean and clear.
To protect the plant before a cicatrice is formed, especially when the surface of the injury is large, we should apply a mastic.

1.2.3 Increase in length

We are aware that there are two types of increase in length.

a) Intercalary elongation:

by an elongation of the cells of this region till they reach their mature state.

b) Terminal elongation:

by three initial superposed cells as in the root, but one forms the epidermis directly,
the second the bark, and
the third the central cylinder.

This elongation ensures regeneration of the terminal bud, which opens out, separating its "leaves" progressively.

It may be noted that gymnosperms have only one initial cell.

1.2.4 Branching of the stem

Branching of the stem begins from buds.

The origin of this branching is external, as opposed to that of branches in the root.[9]

9 It is easy to strip a young branch from a trunk: in debudding of pear or apple trees superfluous branches are removed by hand.
It is difficult to strip a smaller root from the main root.

1.3 Anatomy of a leaf

There is no secondary structure in a leaf.

Leaves are:

— either deciduous, and fall every year,

— or persistent, and fall less regularly, remain longer on the tree and are replaced before they actually drop off.

On the other hand, anatomically, a leaf has two distinct parts: the lamina and the petiole.

1.3.1 Structure of the lamina

Let us examine a transverse section of the lamina of a leaf from a bean plant (*Phaseolus vulgaris*).

From top to bottom, we observe:

a) A layer of flattened cells where the external membrane is slightly cutinised: *this is the upper epidermis.*

b) A tissue composed of connected cells in the form of a palisade, rich in chloroplasts: *this is the palisade parenchyma.*
c) A tissue composed of irregular, rounded cells, with large intercellular spaces and fairly rich in chloroplasts: *this is the spongy parenchyma.*
d) A layer of cells on the lower side that is much less cutinised than on the upper one but with many orifices: *this is the lower epidermis with apertures of the stomata.*
e) In the spongy parenchyma we observe vascular bundles that are simply *veins.*
Secondary veins reinforced with collenchyma or sclerenchyma conduct sap and supportive tissue.
f) The *stomata* that we have seen in the lower epidermis are composed of two bean-shaped guard cells. The ends of these are in contact, leaving a small orifice between them: *the stomatal aperture.*

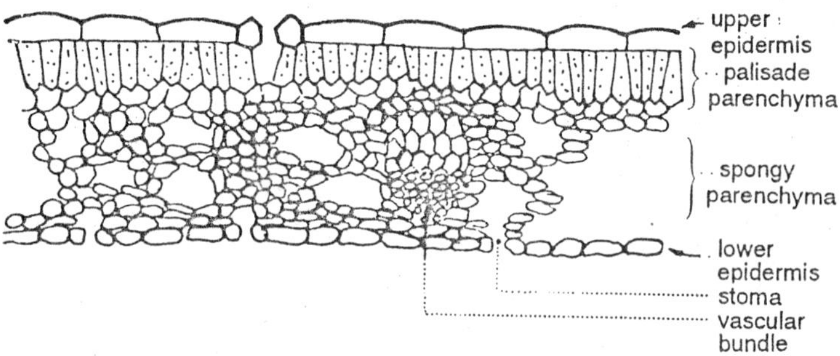

Fig. 4.6. Structure of the lamina of a leaf

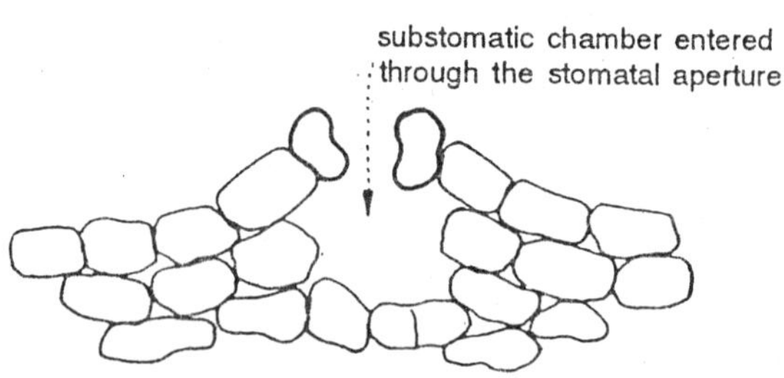

Fig. 4.7. A stoma

This aperture opens into the *substomatic chamber.* Variations in turgidity of the two guard cells of the aperture open or close a stoma.
We observe stomata in places where the substomatic chamber is spongy: this is because of a *water stoma*

or hydathode that releases water in a liquid and non-gaseous form.
We observe several thousands of stomata per square centimetre (see their role in the chapter on Physiology).
We note that there is a bilateral system in respect of the central venation.
The leaf in monocotyledons is considered as a leaf of a dicotyledon folded in two. As a result, we find a lacunar tissue and equal numbers of stomata on both sides.
This is normal because the leaves of all plants in this group are in a position that is near the vertical.

1.3.2 Structure of the petiole

This is actually a simple "conduit" that also has a supportive role.
A microscopic section reveals:

— an epidermis with a few stomata,

— ground tissue of a greenish colour due to chloroplasts,

— a fairly large number of vascular bundles.

If we conduct a microscopic study of a fallen leaf in the autumn we will find a generative layer at the base of the petiole.
It is this layer that regulates *leaf fall.*
At the end of autumn, the layer produces an internal and an external layer of cork perpendicular to the petiole that, later on, forms a gel.

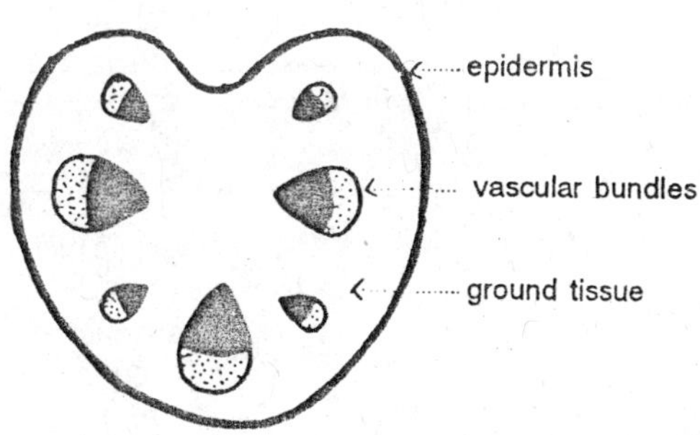

Fig. 4.8. Structure of the petiole

The leaves are retained only by the vascular bundles that surround the petiole: wind causes them to break off.
At the time the leaves fall, the wounds are suberified.

2 Anatomy of the Vegetative Parts of Cryptogams

We will maintain the same sequence of observation for cryptogams as we did during their morphological study, in other words, from the most developed (those closest to phanerogams) to the simplest (unicellular bacteria).

2.1 *Ferns*

Ferns have a root, a stem and leaves that are, anatomically, identical to those of phanerogams.
We observe:
— In the root:
bark,
endodermis,
pericycle, and
phloem and xylem strands with an initial cell for elongation.
— In the stem:
epidermis with stomata,
bark,
endodermis,
pericycle,
but a ring of phloem surrounding the central xylem strand with the initial cell for elongation.
— In the leaf:
lower and upper epidermis with stomata,
parenchyma, and
venation and vascular strands.

2.2 *Mosses: sphagnum*

These begin to differentiate themselves clearly from phanerogams:
a) By an absence of vascular tissues.
b) By an absence of roots.
c) Stems and leaves of sphagnum contain many empty cells, filled with air or water, and with reinforcements of the walls to increase the rigidity of the organs.[10]
Those cavities "transform them" into "sponges":
d) The stems have a bark and pith.
e) The leaves are formed of a layer of cells with venations and respiratory chambers.

10 Sphagnum is used by horticulturists in the preparation of some soil mixtures that are required to preserve a relative humidity.

2.3 Fungi

These are plants without vascular tissue, roots, stems or leaves.

They are formed of a single cell (beer yeast) or several cells: *a thallus.*

A thallus may be septate—edible mushroom (*Psalliota hortensis*)—or non-septate—mildew on the potato (*Phytophthora infestans*).

In the latter, the mycelium has a single cavity with several nuclei (coenocyte).

2.4 Algae

These are anatomically similar to fungi.

Like fungi they have no vascular tissue, roots, stems or leaves.

They may be composed of one or several microscopic or giant cells.

2.5 Lichens

A microscopic section reveals:

— an upper epidermis,
— a filamentary thallus comprising green cells,
— a lower epidermis,
— holdfasts for attachment.

The green cells are unicellular green algae.

The remainder is composed of a fungus.

This is a so-called symbiotic association (*see "Physiology"*).

Chapter 5

Plant Physiology

1

Physiology of Green Plants or Plants with Chlorophyll

We will, in this study, examine a progression:

— taking into account the processes in the development of saps,

— by resolving physiological problems that arise in the course of our work.

1.1 Water and mineral salts

1.1.1 What is the composition of plants?

Assessment of the various elements in the composition of plants is possible by analysis.

1.1.1.1 Analytical method

a) The predominant and essential element is *water*, which usually forms 70 to 90% of a plant and even 95% in fleshy fruits and some leafy vegetables.
On the other hand, this percentage may be as low as 10% in seeds and dry fruits.
The water content is determined by desiccation:
Fresh weight – dry weight = weight of water content.
We are here referring only to *free water*, in other words, the water in the composition of protoplasm to form the "gel" and water in the vacuoles that ensures the turgidity of cells.
This does not include the *water of constitution* of some chemical molecules.
b) Next, after incineration there remains 2 g/l of mineral salts, among which there are:
potassium, magnesium, calcium, zinc, boron, sulphur, iron, phosphorus, arsenic, chlorine, copper, etc.

Albuminoid materials containing carbon, hydrogen, oxygen and nitrogen have, obviously, been destroyed by the fire and released in the forms of carbon dioxide, water vapour and gaseous ammonia.

c) To summarise, a plant usually contains: 45% of carbon, 40% of oxygen and 6% of hydrogen, the remainder being mineral matter, including 3% of nitrogen and 2% of calcium.

This method of analysis helps us to determine the overall composition of a plant but, unfortunately, it does not clearly indicate the role of each of these elements, especially of the last 5%, composed of mineral matter.

Investigators have succeeded in doing so using:

1.1.1.2 The synthetic method

a) Plants are grown in a very dilute nutritious solution and the composition is varied from time to time.

In this manner we are able to ensure the most useful equilibrium.

Knop had developed a formula containing calcium nitrate, potassium nitrate, potassium dihydrophosphate, ferrous phosphate, magnesium sulphate, etc., which involves, in all, the use of 10 chemical elements:

— including carbon from the air (see "Chlorophyll assimilation")

— hydrogen, oxygen, nitrogen, calcium, potassium, sulphur, magnesium, phosphate and iron.

b) Investigators have therefore concluded that:

— mineral salts are essential for plant life,

— they dissolve in water,

— they form mineral solutions[1] in various concentrations, and the essential elements, apart from C, H, O and N, the bases of what are called tertiary and quaternary compounds, are:

K in the form of potassium salts, necessary for bulbs and fruits,

Ca in calcium salts,

S in sulphates found in proteins,

Fe essential for the formation of chlorophyll,

P of phosphates found in organic material,

Mg entering into the composition of chlorophyll,

N observed in the form of nitrates and salts of ammonia.

c) We should, however, take into account the role of secondary elements, such as:

zinc, boron, iodine, copper, molybdenum, silicon and bromine.

1 We will see that C is absorbed from the air; H and O come from water. In agronomical practices three salts have attracted special attention as those that act in large quantities: these are salts of N, P and K.
We are, however, also aware that the non-availability of some other salts may result in defects known as physiological deficiency diseases:
e.g. boron: disease of the cork in apple trees and the heart of the beetroot;
e.g. iron: chlorosis.

1.1.2 Proof of absorption of water and mineral salts

1.1.2.1 If we uproot a plant and leave it exposed to the air, it dies. If we soak it in a nutritive solution, it lives.

Conclusion: A plant receives nutrition through its roots.[2]

2

a) We must pay attention to the root system of a plant whenever it is planted or transplanted.
b) If we are providing a plant with fertiliser this must be placed as close as possible to the absorbent root hairs and not, for example, at the base of the trunk.

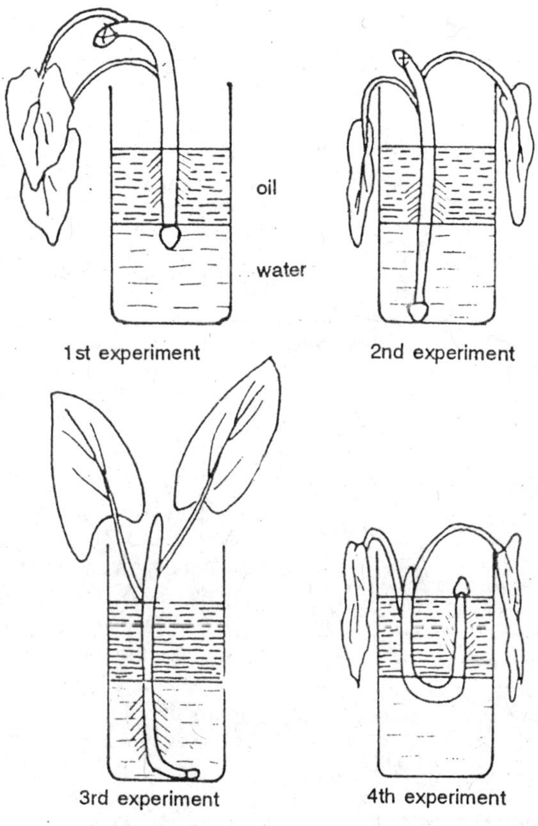

Fig. 5.1. A plant absorbs nutrition through its absorbent root hairs

1.1.3 Mechanism of absorption

1.1.3.1 Let us repeat the Dutrochet experiment

We take a trough filled with pure water.
In this trough we place a glass jar fitted with a tube and the broader portion closed with a bladder, a permeable membrane.

The glass jar is filled with a sugar solution. The two levels are in equilibrium.
Initially, the volume of liquid in the glass jar increases and the level rises in the tube. Pure water passes through the permeable membrane.
Subsequently, the level falls and an equilibrium is again established, but now the pure water in the trough is sweetened. There has been a passage of sugar through the permeable membrane.

3
This osmotic pressure induces a turgidity in the cell. This phenomenon, repeated in every cell, holds the plant erect.
When there is a shortage of water in the soil the osmotic pressure is reduced and the plant "wilts" or "droops".

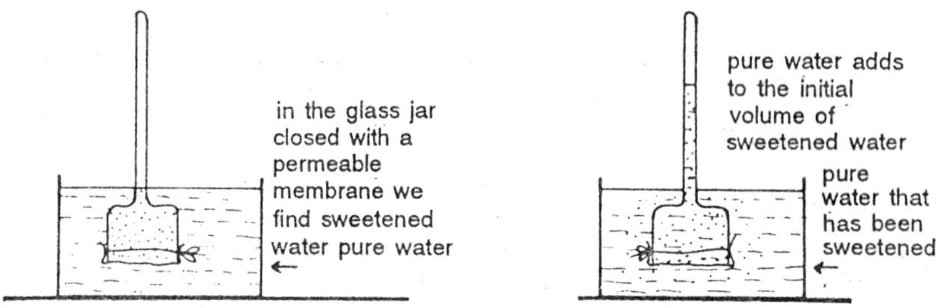

Before experiment | After experiment (Ist reading)

Fig. 5.2. The Dutrochet experiment

Conclusions
1) Crystalloid substances (such as mineral salts) may pass through a permeable membrane and the cellulose wall of a cell is such a membrane.
2) There is an attempt to attain an equilibrium between the two solutions separated by a permeable membrane. The less concentrated solution is, initially, attracted by the more concentrated one and this results in an increase in pressure in the container of the concentrated solution.
This pressure is called *osmotic pressure*[3] and is higher when there is a considerable lack of equilibrium.
3) Only crystalloid solutions can pass through a permeable membrane, not colloidal solutions. This explains why the contents of a cell cannot escape.
4) A cellulose wall can, therefore, be penetrated by water and various mineral salts, but a plasma membrane selects some of the required salts.
This selection is regulated on the basis of use.
The more a salt is used by a plant, the easier for the salt to enter it, and vice versa.
Unfortunately, some harmful salts are not "selected": e.g. sodium chlorate.
5) This mechanism, called osmosis,[4] helps in absorption, through the root hairs, of water and mineral salts which, in this way, reduce the concentration

4
a) This phenomenon, in view of a low concentration of mineral salts in the soil and, therefore, a low concentration of minerals available for plants, explains why a large amount of water must be made available and it is for this reason that fertilisers are called "water savers".
b) The mechanism for absorption of mineral salts also explains why the root system is larger when the soil is poor in mineral elements and water,
c) The selectivity of roots in respect of various mineral salts, considering the requirements of a particular species, partly explains the necessity for the rotation of crops as well as the requirements of certain plants: acid soils, basic soils.

of vacuolar sap, whence arises a requirement of an adjacent cell, and so on: there is a penetration from cell to cell.

1.1.3.2 Though roots absorb mineral matter by osmosis they also encounter insoluble substances: hard rock, cement, etc.

Roots emit acids[5] (carbonic acid, oxalic acid), the role of which is to dissolve these substances.

1.1.4 Role of roots

The role of roots is one of fixation as well as of absorption.
Their development is proportional to the richness of the soil in fertilising elements but is often similar to that of aerial organs in normal conditions of growth.[6]

1.1.5 What is raw sap?

Mineral salts in aqueous solution form raw sap.

1.2 Transpiration

1.2.1 Definition

This is the way in which a plant releases an excess of water in the raw sap into the atmosphere as water vapour, thus maintaining its concentration.

1.2.2 Proof of transpiration

1.2.2.1 Let us take a plant potted in a varnished container, close it hermetically above and below and place it in a sealed bell jar.

The inside of the bell jar is seen to be covered with moisture.
Conclusion: A plant transpires through its aerial organs.

1.2.2.2 The same plant, in the same container, is placed on a weighing machine.

After some time there is a loss of weight.
Conclusion: A plant certainly transpires through its aerial organs and the difference in weight recorded in the aforementioned experiment indicates the extent of such transpiration.

5

These secretions being specific to each plant species, this partly explains the reasons for crop rotation because some excesses in the quantities of the formed products may even contaminate the soil.

6

The horticulturist tries by all means to assist in development of the root system:

- **by helping in growth: ploughing, draining . . .**
- **by helping in multiplication:**
- **by replanting: development of fasciculate systems.**
- **by ridging**
- **by rolling**

} development of adventitious roots.

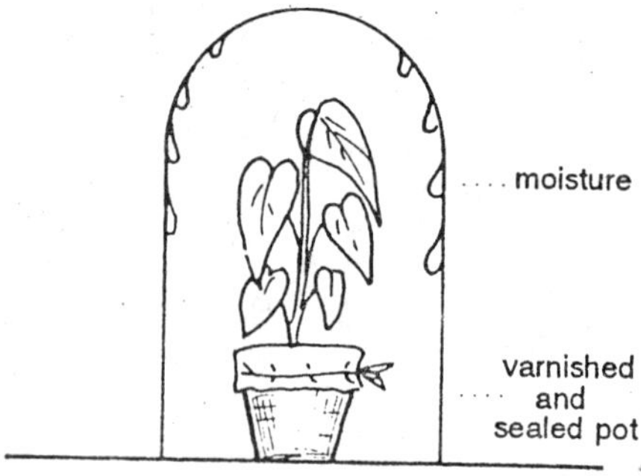

Fig. 5.3.

1.2.2.3 Potometer experiment

A U-tube has a plant at one side and, at the other, a thin tube filled with water that is hermetically sealed at the base of the leaves of the plant.
After some time, the water in the thin tube is displaced. Its volume has been reduced.
Conclusion: It is the roots that absorb water and the leaves that transpire.

1.2.3 Organ for transpiration

We know that it is leaves that transpire.
Let us place a tube containing calcium chloride and hermetically sealed on both surfaces of a leaf.

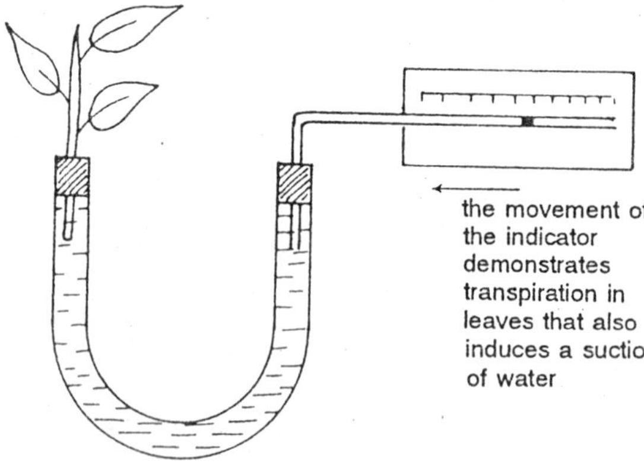

Fig. 5.4. Potometer experiment

From the difference in weight, we note that the lower surface of the leaf has released more water

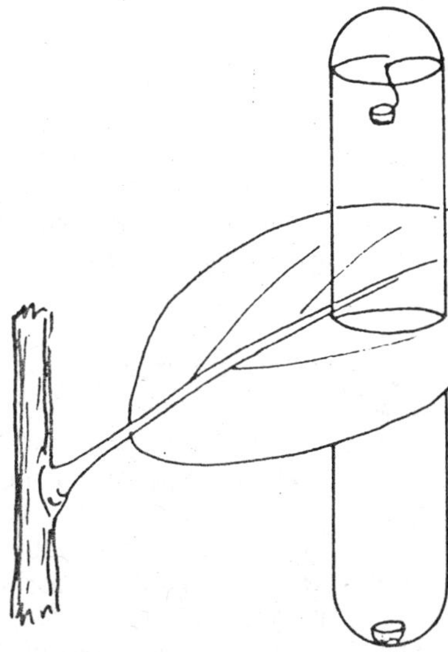

Fig. 5.5. Leaves transpire through their stomata (more numerous on the lower surface of leaves)

vapour (heavier tube: the calcium chloride has fixed a lot of water) and the upper surface has released less.
Conclusion: The lower surface has many stomata.
The upper surface transpires only through its cuticle.
Leaves, therefore, transpire mainly through their stomata

1.2.4 Rate of transpiration

1.2.4.1 Volume of water transpired

The volume of water transpired may be considerable.
Some figures are:

—from 100 litres per day for an apple tree,
—the same amount for a beech tree,
—1,200,000 litres for a hectare of meadow per year,[7]
—3,000,000 litres for a hectare of wheat per year.

1.2.4.2 Variations in transpiration

These depend on:
a) The leaf surface, in other words on the number of stomata.[8]

7

a) Since this volume of water is needed to conduct mineral salts in the raw sap to the leaves we may assume that in all horticultural plantations we must economise on water: store it by deep ploughing and prevent its evaporation by harrowing whenever rainfall is inadequate.
b) It should be borne in mind that any uprooting of trees results in:
—a reduction in atmospheric water vapour
—an increase in soil humidity that may cause waterlogging.
Some humid soils can be corrected by planting poplars that draw up a lot of water.

8

If a plant transpires much but does not have plenty of water, it wilts.
This applies to:
—cuttings with leaves on which we must, therefore, reduce the foliar surface;
—transplants, of which the aerial portion must be less than the underground portion.

b) Temperature:[9] maximum at 45°C. Beyond this temperature a plant will die.
c) The amount of moisture in the air.[10]
d) The movement of air: moving air increases the rate of transpiration.
e) Light: which causes the stomata to open.
f) Time of day: maximum at 11 o'clock.
g) The plant: herbaceous plants transpire more than woody plants.

9
When The average temperature in a plantation is high, more frequent and copious irrigation and sprayings are required.

10
Self-resistance in some plants is ensured:
— by a cutinisation of their leaves,
— by a transformation of leaves into thorns (cacti).

1.2.5 Sudation

This is the elimination of excess water from raw sap in liquid form and not as vapour.
Sudation or guttation completes or replaces transpiration when external conditions limit or prevent it: as in the case of excessive atmospheric humidity.
This is proof that excess water must be eliminated.
It is manifest in droplets of water that are formed along the margins and ends of leaves.
This takes place through water stomata or hydathodes and when the aquiferous tissue is filled with sugar there is a formation of nectar: as in flowers.

1.3 Chlorophyll assimilation

1.3.1 Definition

1.3.1.1 Carbon, vital element for a plant

We know that plants absorb water and mineral salts from the soil.
Carbon exists in three forms.
a) Organic carbon, but green plants cannot assimilate it.
b) Mineral carbon, the same problem, because carbonates are not usually soluble.
c) This leaves us with so-called atmospheric carbon that combines with oxygen to form carbon dioxide.
Green plants, therefore, fix carbon from carbon dioxide in the air (or dissolved in water, by aquatic plants: *Elodea canadensis*).

1.3.1.2 What is chlorophyll assimilation?

Chlorophyll assimilation is the function whereby green plants, when exposed to light:
1) absorb carbon dioxide from the air,
2) release oxygen,

3) synthesise elements of elaborated sap by a combination of those of raw sap and carbon dioxide.

1.3.2 Proof of chlorophyll assimilation

1.3.2.1 Green terrestrial plants

Let us put a plant under a bell jar with a certain volume of carbon dioxide.
After a while, analysis shows that:

— the amount of carbon dioxide has been reduced,

— the amount of oxygen has increased while the plant was exposed to light, whereas there was no increase in the oxygen level in darkness (*see* §*1.4*).

Conclusion: When exposed to light, a green plant absorbs carbon dioxide and releases oxygen.

1.3.2.2 Green aquatic plants

Let us carry out the same experiment with a well-known plant from an aquarium: *Myriophyllum hippuroides* (or any other aquatic plant) placed in an inverted funnel, covered with a test tube.
If the water is rich in carbon dioxide, gas is released in bubbles that collect at the base of the test tube in the presence of light.
If we place an almost extinguished match in the tube, after upturning it, there is a bright flame: it is, therefore, oxygen that is released.
If the water does not contain dissolved carbon dioxide there will be no release of gas.

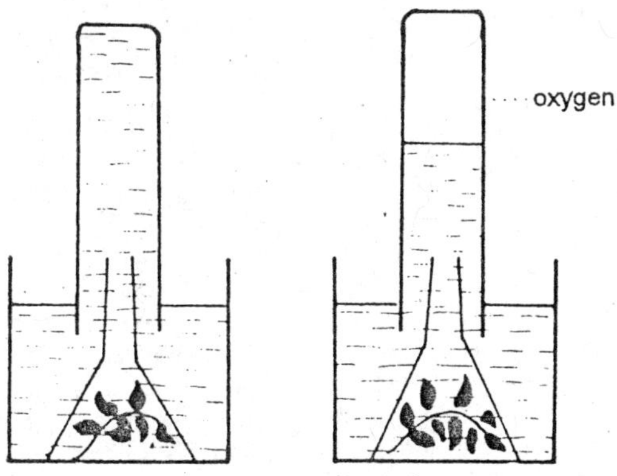

Fig. 5.6. At the beginning of the experiment (left) and at the end of the experiment (right)

1.3.3 Rate/influence of chlorophyll assimilation

1.3.3.1 Amount of chlorophyll assimilation

We have noted that this happens only in the presence of light.
During this time, respiration continues and the function bears an inverse relationship to chlorophyll assimilation: carbon dioxide is released and oxygen absorbed in the process.
We proceed to carry out two separate measurements:

— measurement of the two combined functions,
— measurement of respiration by keeping the plant in a dark place.

By the difference between the two measurements we obtain the value of chlorophyll assimilation.
By assimilation quotient we mean the oxygen–carbon dioxide relationship.

1.3.3.2 Variations in chlorophyll assimilation

These depend upon:

a) Temperature
feeble at 0°C,
optimum at 25°C,
maximum at 45–50°C,
beyond this temperature, a plant will die.[11]

b) The concentration of carbon dioxide.
Air has a natural content of 0.03% carbon dioxide, or 300 ppm, which is adequate, but the optimum is about 500 or 1000 ppm depending upon the plant.[12]

c) Light, and especially the duration of light, its intensity and its wavelength.[13]

11
Most plants can develop only at a certain temperature. For this reason horticulturists try to ascertain the optimal temperatures: plantations under cover.

12
Provision of carbon dioxide in greenhouses to the extent of a 10 per 10,000 concentration, i.e. 1000 ppm or even 1500 ppm (ppm = parts per million), increases the intensity of chlorophyll assimilation resulting in a rapid growth and an increase in production.
This technique is used for lettuce (*Lactuca sativa*), cucumbers (*Cucumis sativus*), etc. We understand that this provision is only useful during daytime.

13
We must consider the requirements of plants: plants that grow well in sunlight—(cacti), rose (*Rosa*)—need more light than "semi-shade" or "shade" plants: ferns, cyclamen (*Cyclamen persicum*). Horticulturists assist chlorophyll assimilation by increasing the duration of illumination or its intensity. In this way they hasten the synthesis of elaborated sap and enable the plant to grow more rapidly. This technique is useful for:
— growing lettuce (*Lactuca sativa*) in greenhouses.
— growing plants in greenhouses for propagation purposes.
— the growth of young plants, chrysanthemum (*Chrysanthemum indicum*), cyclamen (*Cyclamen persicum*), saintpaulia (*Saintpaulia ionantha*), gloxina (*Gloxina barbata* . . .).
Nevertheless, this process (action of photosynthesis) has nothing to do with the provision of light to modify the photoperiodism and, thus, act on the cycle of appearance of flowers for a given plant (see Chapter 6, §1.1).

d) Leaf surface.[14]

1.3.4 Mechanism of chlorophyll assimilation

1.3.4.1 Chlorophyll

Chlorophyll assimilation occurs by means of a green pigment: colouring of leaves and stems of herbaceous plants occurs on account of localisation of chlorophyll in chloroplasts.

Chlorophyll is formed only in sunlight.[15]

a) Composition of chlorophyll

Let us collect a few dark-green leaves: spinach (*Spinacia oleracea*) or cherry laurel (*Prunus lauro-cerasus*).

Cut and macerate them for 24 hours in alcohol. Then filter the solution.

Obtain an alcoholic solution of raw sap.

To this add benzene, stir it and leave it to settle.

On the surface of benzene, which has become green, we find *pure chlorophyll.*

Below, in the alcohol that has acquired an orange colour, we find a mixture of yellow *xanthophyll* and red *carotene.*

Only pure chlorophyll plays a role in chlorophyll assimilation. Chemically, chlorophyll is a proteinic substance containing magnesium.

b) Absorption spectrum of pure chlorophyll

Pure chlorophyll absorbs solar radiations to extract energy from light.

But which radiations are used?

When light is broken up by a prism, it reveals a visible spectrum of seven colours:

red, orange, yellow	thermal rays
green, blue, indigo, violet	actinic rays

Let us carry out the following experiment:

We break down solar illumination with a prism. In front of each of these seven bands, we place a tube filled with water, charged with carbon dioxide, turned upside down and containing an aquatic plant.

After some time we note:

— a release of a large amount of oxygen in the "red" tube,

— a release of a moderate amount of oxygen in the "orange" "blue", "indigo" and "violet" tubes,

14 This is why horticulturists are especially careful about the leaf surface of plants even when they are not grown for decorative purposes (for example, fruit-bearing plants).

15 Blanching or yellowing of green tissues in darkness. This technique is used by market gardeners for certain vegetables: Chicory (*Cichorium intybus*), endive (*Cichorium endivia*), leek (*Allium porrum*).

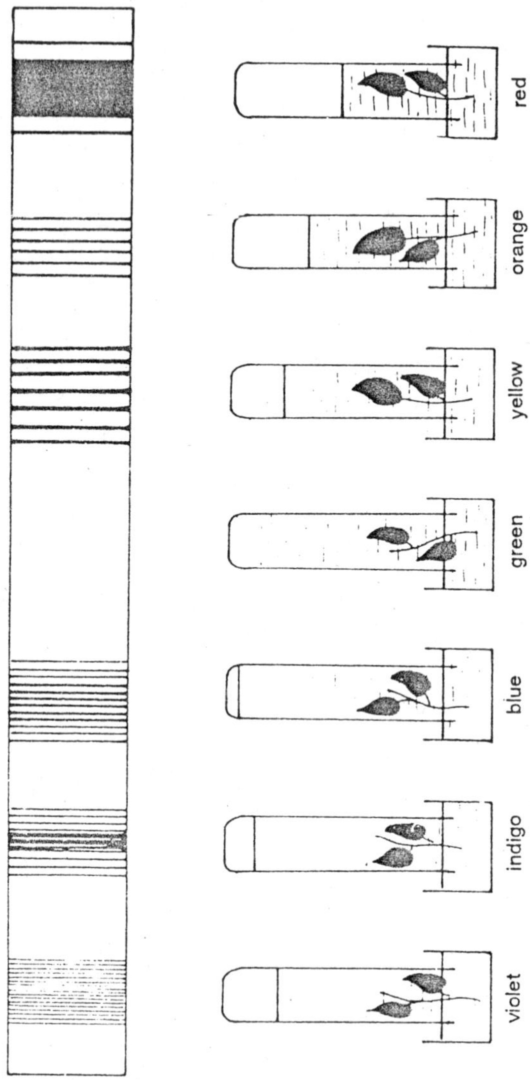

Fig. 5.7. Absorption spectrum

— a minimum release of oxygen in the "yellow" tube,

— no release in the "green" tube (*see Fig. 5.7*).

Conclusion: Chlorophyll does not use green radiations.[16]

Sometimes supernumerary pigments conceal the presence of chlorophyll.

In this connection we may refer to:

— coloured leaves of some ornamental plants,

— algae that contain phycoerythrin (red alga), phycocyanin (blue alga) and phycophein (brown alga).

Phycocyanin displaces the absorption region towards the blue, the only radiation capable of deep penetration in the sea.

16

A plant exposed to green radiations only cannot grow. On the other hand, red and violet radiations help in growth.

1.3.4.2 Chemical products

Based on the latest investigation results of our researchers we can say that:

a) Optical energy which is absorbed enables the break-up of water in raw sap and its separation as hydrogen (H) and oxygen (O).

b) The combination of carbon dioxide (CO_2) absorbed by chlorophyll assimilation and hydrogen from water (H_2O) results in the formation of sugar ($C_6H_{12}O_6$) and water (H_2O).

Conclusion: We can, generally, summarise these early stages as follows:

$$6CO_2 + 12H_2O \longrightarrow \underbrace{C_6H_{12}O_6 + 6H_2O}_{} + 6O_2$$

$6CO_2$	$12H_2O$	$\longrightarrow$	$C_6H_{12}O_6 + 6H_2O$	$6O_2$
by chlorophyll assimilation	from raw sap	reaction assisted by chlorophyll as a "catalyst" that does not itself take part in the reaction	sugar (glucose + water)	released in the atmosphere

All this is achieved with the help of calories provided by light.

c) Glucose is transformed into starch at a rapid rate in daylight.

We may conduct the following experiment:

On a sheet of black paper we cut out some portions. The sheet of paper is placed on a green leaf and removed in the evening.

The lamina is dipped in iodised water; the portions under the cut-outs are indicated by a blue colour because they were the only portions exposed to light and chlorophyll assimilation resulted in the formation of glucose in them that has been transformed to starch (causing iodine to turn blue).

d) At night, starch is again transformed to glucose which is easily transportable towards the places where reserves are stored (*see* § *1.6*).

1.3.5 Role of chlorophyll assimilation

A plant increases in weight due to a fixation of carbon, several tonnes of carbon per year in a hectare of forest.

A plant cleanses the atmosphere by absorption of carbon dioxide, but only in daylight.

The function that enables the synthesis of organic matter with the help of light is also called photosynthesis.[17]

Raw sap, concentrated by transpiration due to the effect of photosynthesis, is transformed into elaborated sap.

1.4 Respiration

1.4.1 Definition

Plant respiration is a function, similar to that in animals, by which a plant absorbs oxygen and releases carbon dioxide by combustion of organic matter.

1.4.2 Proof of respiration

1.4.2.1 Let us conduct the following experiment

Employing:

— either fungi (plants with no chlorophyll) in light,
— or green plants in darkness,
— or green plants under anaesthesia

to neutralise chlorophyll assimilation, the plant is placed under a bell jar with a base containing lime water.

After some time, the lime water becomes milky, much more milky than if the same experiment were carried out without the plant being confined to the bell jar. This proves the release of carbon dioxide which, on contact with lime water, forms a precipitate of calcium carbonate.

1.4.2.2 Let us place some viable seeds in a hermetically sealed jar

After a while, a candle inserted into the jar is extinguished: this is proof of a reduction in the amount of oxygen in it.

---

17 An analysis of plants can enable an assessment of their proper nutrition.
This can be done on an entire plant or on a part of it (for example, on its leaves).
It is of economic interest in the case of perennial plants, providing a better understanding of their nutrition in N-P-K and microelements enabling them to continue from one growth cycle to another (as in fruit arboriculture).
In horticulture (on hortensia and pelargonium) it will help to draw conclusions on the effects of adopted nutritive solutions.
Actually, we are dealing with an improved fertilisation by refining the results of soil analysis. The surest method is by a comparison of samples, some taken from plants "in a good state of growth" and others from plants that have "created problems", provided the samples are selected at the same stage of growth.

Fig. 5.8. Plants respire

Conclusion: All plants respire and, moreover, all organs respire.

1.4.3 Rate/influence of respiration

Respiration varies depending on:

a) Temperature

Most of our indigenous plants have no respiration at about 0°C and a maximum at 45-50°C.

Above this temperature a plant will die.

On the other hand, a fir tree (*Abies alba*) in cold regions survives even at −20°C.[18]

b) Age and period of development

The maximum rates of respiration are recorded during periods of high activity.

Herbaceous plant: germination and flowering.

Tree: budding and flowering.

c) The organ

All organs respire: roots, stems, leaves, reproductive organs, seeds ....[19]

d) The plant

An annual plant respires more than others.

1.4.4 Chemical products of respiration

Respiration is an oxidation (slow burning) which destroys organic matter to form other substances that are necessary and more assimilable, by restoring the basic compounds: carbon dioxide—CO_2, water—H_2O, and ammonia—NH_3.

18
Since respiration is accompanied by a loss of carbon (released carbon dioxide) involving a loss of weight, the horticulturist will find it useful to limit this function. This is also, and particularly, another reason (see Chapter 6, §1.1.2) for limiting the temperature at night in greenhouses.

19
Roots respire and, therefore, a plant should be grown only in a soil that is aerated: by digging, draining . . .
Since the main exchanges are through the leaves, these must be cleaned regularly when a plant is kept in a dusty environment.

We may summarise these reactions as follows:

$$C_6H_{12}O_6 + 6O_2 \longrightarrow 6CO_2 + 6H_2O + \text{Energy of plant heat}$$

glucose formed by chlorophyll assimilation	oxygen absorbed by respiration	carbon dioxide released by respiration	water

This is a reaction opposite to that of chlorophyll assimilation.

Conclusion:

1) While chlorophyll assimilation synthesises carbonaceous material, respiration destroys it, but much less than what is formed by chlorophyll assimilation.

2) Whereas chlorophyll assimilation consumes the energy that it extracts from light radiation with the help of chlorophyll, respiration generates energy.

The energy generated is partly used for synthesising substances necessary for a plant.

The remainder is released in the form of plant heat: seeds when piled up release a certain amount of heat even when they are dormant.[20]

20 Viable seeds to be used for germination are not to be kept in silos. The rise in temperature may be excessive and harmful for the embryos in these seeds.

1.4.5 Respiration and asphyxiation

1.4.5.1 When a plant is kept in an atmosphere that has a low level of oxygen, it resists asphyxiation by using its reserve materials

Some materials, like sugar, can, by decomposing, liberate energy that replaces the energy produced by respiration.

It may be noted that this total energy is less in anaerobic respiration than in aerobic respiration and, if the struggle against asphyxiation continues for long, alcohol formation becomes excessive and the plant dies.[21]

21 This is also what happens in the bletting of fruits that occurs due to asphyxiation (because fruits respire) with the formation of alcohol from glucose.
There should not, however, be any confusion between bletting that begins in the centre of the fruit, and rotting that begins from the outside.

1.4.5.2 Alcohol is formed by decomposition of glucose. This is the very principle of alcoholic fermentation.

Alcohol is formed at the cost of glucose, because beer yeasts that have been activated and kept in anaerobic conditions resist asphyxiation.

There are many types of fermentation (lactic, acetic, etc).

1.5 Circulation of saps

1.5.1 Raw sap

1.5.1.1 Verification of raw sap circulation

a) We cut a plant at ground level.
On the cut portion that remains in the ground we observe the appearance of droplets, proof of the rise of sap from the roots.
b) We put a red dye in the water where a part of the plant is placed: soon we observe a reddening of the xylem vessels.
Conclusion: Raw sap circulates from the roots towards the leaves.

1.5.1.2 Causes for the rise of raw sap in xylem

Three causes have been advanced, each playing a more or less important role.
a) Osmotic pressure
Hales conducted the following experiment:

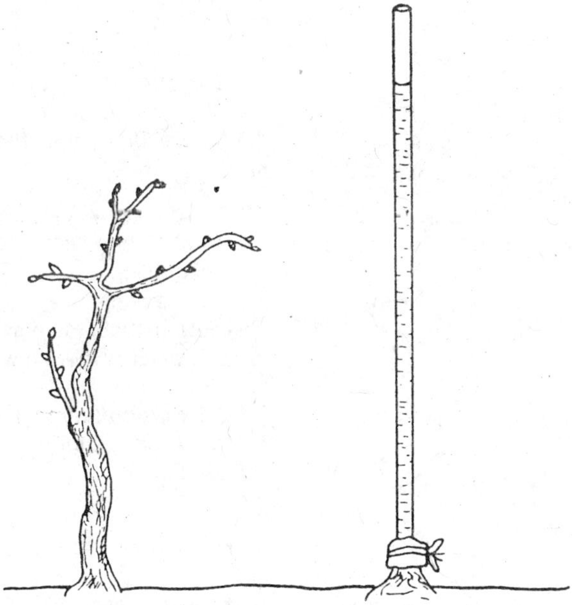

Fig. 5.9. Hales' experiment indicating osmotic pressure

At the top of a vinestock (*Vitis vinifera*) that was cut, he attached a mercury pressure gauge (manometer). Result: the pressure recorded could, sometimes, be as high as two atmospheres.

This is explained by the phenomenon of osmosis that enables a plant to draw water and mineral salts from the soil.

The roots then act as force pumps, *a very important role* in the rise of sap.

b) Transpiration

The release, in the form of vapour, of excess water contained in raw sap creates a continuous suction from the bottom upwards.

The leaves act as a suction pump but, as such, *have only a limited role* because purely physical considerations oblige us to limit these possibilities of action to a height of about 10 metres.

c) Capillarity

This physical phenomenon that causes liquids to rise in tubes of small diameter appears to have *only a limited role* because, when the xylem elements are very small, a certain number of air bubbles circulate with the raw sap and, blocking the liquid column, could interrupt the capillarity.

1.5.2 Elaborated sap

1.5.2.1 Verification of the circulation of elaborated sap

We impregnate an incision on a leaf with fluorescein. Examination of a section of the stem, in light, reveals the penetration of this fluorescent material into the phloem elements.

Conclusion: Elaborated sap circulates from the leaves towards places where it will be used.

1.5.2.2 Elaborated sap is a viscous liquid, poor in water and mineral salts and rich in organic substances

It is:

a) used up on the spot as an element of growth

or

b) stored in various organs to be kept as a reserve for future use, or accumulated in what are called secretory tissues and, in principle, not used by the plant.

1.6 Nutritive reserves in plants

1.6.1 Localisation of reserve materials

All the organs of higher plants can store nutritive reserves.

These are, briefly:

■ *Roots* that have become tuberous: carrot (*Daucus carota*), turnip (*Brassica napus*), beetroot (*Beta vulgaris*), dahlia (*Dahlia hortensis*).

■ *Stems*

Aerial stems: sugarcane (*Saccharum officinale*), sugar maple (*Acer saccharum*), palms.

Underground stems: potato tuber (*Solanum tuberosum*), gladiolus bulb (*Gladiolus*), rhizome of couch-grass (*Agropyron repens*).

■ *Leaves*

Aerial: sempervivum (*Sempervivum arachnoideum*).

Underground: onion (*Allium cepa*).

■ *Fleshy fruits*

Apple (*Malus communis*).

■ But, most often, nutritive reserves are located in *seeds*, either in the endosperm or in the cotyledons: as seen in cereals, dry vegetables, and oil plants.[22]

22 So-called food plants are grown by man to use the accumulated reserves.

1.6.2 Principal reserve materials

1.6.2.1 Carbohydrates contain

Carbon }
Hydrogen } CHO
Oxygen }

a) Starch

This is the most common reserve material. It is of the "storage product" type because it is stable. Starch does not circulate in elaborated sap.

It is most frequently found:

— in seeds—peas (*Pisum sativum*), beans (*Phaseolus vulgaris*)—to the extent of half their weight.

— in underground organs: the potato tuber (*Solanum tuberosum*), tapioca of manioc (*Manioc utilissima*).

Under the microscope starch is seen in the form of ovoid grains, in concentric layers within the cell.

From its various forms and sizes we can easily identify its origin.[23]

Starch can be coloured blue with iodine.[24]

23 This identification helps in preventing frauds when flour is adulterated.

24 Its easy detection can, for example, identify ripe fruits ... which should not contain starch (see Chapter 6, §1.5.3).

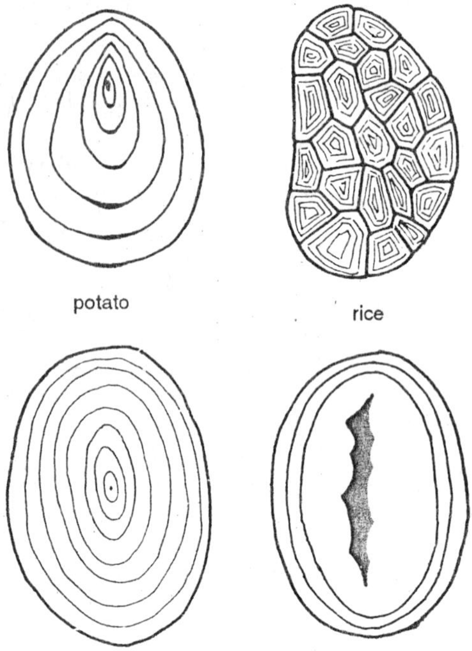

Fig. 5.10. Various forms of starch in different plants

b) Inulin

This is a material similar to starch but it is soluble.[25]
It is formed by a polymerisation of levulose.
It is found in:

— the Jerusalem artichoke (*Helianthus tuberosus*);
— the salsify (*Trapopogon porrifolius*);
— the globe artichoke (*Cynara scolymus*).

c) Sugars

Carbohydrates used by the plant are always in this form.
They also exist in the form of sugars that circulate in elaborated sap.[26]
Sugar exists in two principal forms:
"fruit sugars" or *glucose* and *dextrose*:
They are directly assimilable. These carbohydrates can ferment and produce alcohol.
Crystallised dextrose forms the bloom on fruits.
"cane sugar" or *sucrose*
This must be transformed into "glucose" to be assimilated: it then provides sugars of the glucose, dextrose or levulose (fructose) types.

25 Its solubility explains why artichokes cannot be stored in the open air like potatoes. They have to be uprooted as and when required.

26 This explains the taste and food value of certain fleshy fruits: pear (*Pyrus communis*), apple (*Malus communis*), peach (*Persica vulgaris*), cherry (*Cerasus avium*), plum (*Prunus domestica*).

We find sucrose in the roots of the beet, in fruits with glucoses and in the nectar of flowers.

1.6.2.2 Fatty materials or lipids contain C, H, O and sometimes phosphorus (P)

They are found in three forms:
a) Combined with phosphorus they form phospholipids, components of the cell.
b) In the form of wax they have a protective role.
c) In the form of fats they are reserve materials.
They accumulate
— in oil seeds: turnip (*Brassica napus oleifera*), rapeseed (*Brassica campestris oleifera*), flax (*Linum usitatissium*), walnut (*Juglans regia*), groundnut (*Arachis hypogaea*);
— in oil fruits: olives (*Olea europaea*).
The edible fat *"vegetaline"* is manufactured from the "butter" of the coconut (*Cocos nucifera*).
Lipids circulate in elaborated sap in the form of fatty acids and glycerine.

1.6.2.3 Nitrogenous materials or proteins contain C, H, O + nitrogen (N), sometimes (P) and sulphur (S)

a) Aleurone
Proteins, in "storage" exist in the form of grains known as aleurone grains that are formed from dried vacuolar saps. They are found in the seeds of cereals, leguminous plants and oil plants.
Aleurone from cereals is called gluten.
Aleurone is usually combined with starch.
b) Amino acids
Asparagine from asparagus (*Asparagus officinalis*), for example.
It is in this form that proteins circulate in elaborated sap.

1.6.3 Utilisation of reserves, enzymes, vitamins and hormones

We know that a plant uses its reserves:
— to grow (elongation . . .) and
— to form new organs (from seeds . . .).
When used, these materials are transformed, liquefied and transported by the phloem, towards places where they are to be utilised.
These transformations can occur only because of enzymes.

Thus, in:	the enzyme	transforms	into
carbo- hydrates	sucrase invertase	sucrose sucrose	glucose + levulose
	amylase	starch	glucose + fructose dextrose
lipids	lipase	oils	fatty acids
proteins	protease	proteins	amino acids

a) Enzymes are biological catalysts, composed of proteinaceous substances that are effective in very small amounts.
Their role is to digest reserve materials and thereby ensure their transformation.
There is one of these per category of substances.
This explains how carnivorous plants digest trapped insects, as in the case of the nepenthes (*Nepenthes alata*), drosera (*Drosera intermedia*).

b) Vitamins
Often combined with enzymes, vitamins are found in coloured fruits and green leaves.
They are essential for both animal and plant life.
It may be noted that they are destroyed at a temperature of 100°C.

c) Hormones (see *§1.8)*.

1.7 Substances not used by a plant

These substances are also stored in a plant but their role is not clearly understood.
Until now they were considered to be waste products eliminated by plants.
Many are used by man.
We will mention some of them.

1.7.1 Mineral substances

a) In the form of an oxalate of calcium: needles in onion cells (*Allium cepa*), crystals in cells of the begonia (*Begonia gracilis*).

b) In the form of calcium carbonate: leaf of the fig tree (*Ficus carica*).

1.7.2 Organic substances

1.7.2.1 Essential oils

These are volatile carbides of hydrogen, soluble in alcohol, that yield "perfumes";
- *in leaves:* thyme (*Thymus vulgaris*), lavender (*Lavandula spica*), mint (*Mentha viridis*);
- *in petals (epidermis)*: rose (*Rosa*), lemon (*Citrus limon*), orange (*Citrus sinensis*);
- *in canals:* pine (*Pinus sylvestris*).

1.7.2.2 Resins

Mixtures of essential oils and oxidised products of essential oils: resin of conifers; copal. Colophony is a resin from which the essential oil has evaporated.

1.7.2.3 Tannin

This prevents putrescence in hides and skins; it also reduces bletting in fruits.
- vacuoles in the oak (*Quercus pedunculata*)
- unripe fruits

1.7.2.4 Alkaloids

These are organic products.
Bases (alkalis) in plants combine with acids producing salts.
Many are fairly toxic in strong doses:
- caffeine, in the seed of the coffee plant (*Coffea arabica*),
- morphine (from opium), fruit of the poppy (*Papaver somniferum*),
- quinine, bark of South American trees,
- nicotine, tobacco leaf (*Nicotiana tabacum*),
- cocaine, coca leaf,
- digitalin, digitalis (*Digitalis purpurea*), etc.

1.7.2.5 Latex

Emulsions containing sugar, starch, resin, protein, etc.
Latex can coagulate. We find it in:
- the dandelion (*Taraxacum officinale*), white latex;
- hevea (*Hevea brasiliensis*);
- India rubber tree (*Ficus elastica*), rubber;
- celandine (*Chelidonium* majus), yellow latex;
- poppy (*Papaver somniferum*), a latex that yields opium.

1.7.2.6 Glucosides

Mixtures of glucose with another organic substance.

— Amygdalin in the fruits of the almond (*Prunus amygdalus*).

— Sinigrin in mustard (*Sinapis alba*).

1.8 Hormones

We are now aware that certain factors regulate the growth of green plants. These are:

— *anatomical factors*: existence of multiplicative centres (meristems),

— *external factors* that intervene in all physiological problems studied in this section.

We must, however, take into account the effect of certain internal substances.[27]

27

We will, briefly, enumerate the principal new techniques that are currently adopted, based on the action of hormones, leaving it to technical courses to explain these in detail. We will not refer to the effect of gibberillins still at an experimental stage.

a) New distributions of natural hormones: bending of fruit trees.

b) Effect of synthetic hormones.

i) Effect by proliferation:

- with a normal dose: stem cuttings, rapid development of roots, budding of fruits, obtaining parthenocarpic fruits, and their early maturity,
- in strong doses: selective or total weed killing.

ii) Effect due to slowing down the rate of growth.

- thinning of fruits,
- delay in ripening of fruits,
- fall of fruits.

iii) Extending the duration of dormancy:

- forcing: delayed flowering,
- antigerms in the potato (*Solanum tuberosum*),
- delaying budding in fruit trees by 15 days in the spring to prevent damage due to frost.

iv) Reducing the period of dormancy:

—forcing: hastening flowering.

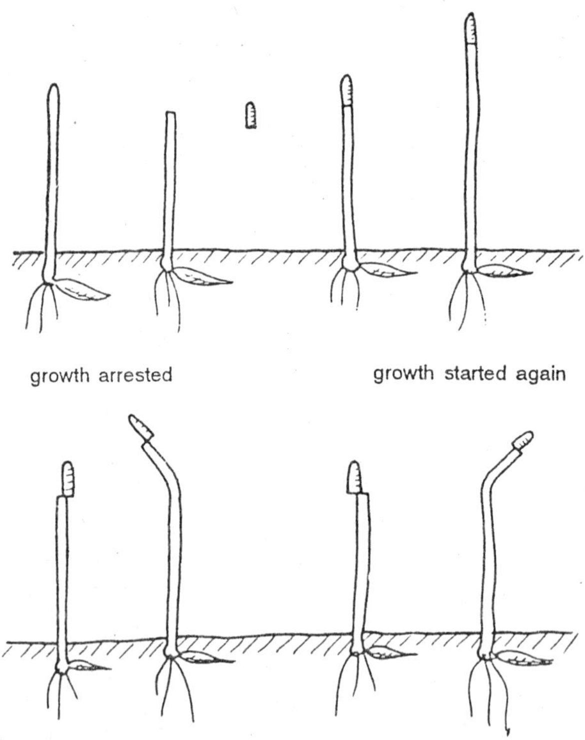

Fig. 5.11. Only the side placed on the coleoptile grows

1.8.1 Discovery of hormones

Careful investigation of the causes of phototropism

(*see* §*3.2.2.3*) and experiments with the oat (*Avena sativa*) have enabled this discovery.

The "coleoptile" of the oat is an organ that protects young leaves and the stem.

If we cut the top of a coleoptile there is no more growth. If we put it back, growth begins again. If displaced to one side or another the elongation is displaced because only the side below the coleoptile appears to grow.

Conclusion: The tip of the coleoptile releases a growth-promoting substance comparable in all ways to animal hormones and known as auxin.

1.8.2 Natural and synthetic auxin

On analysis, natural auxin when discovered was found to be indol β acetic acid.

Other synthetic compounds showed the same functions:

— naphthalene α acetic acid

— indol butyric acid.

1.8.3 Action of auxins

Auxins are characterised by an effect on:

— cell multiplication and

— cell elongation.

They do not rise; their effect is downwards and from a distance.

They have an effect on all the organs of a plant, but only in very small doses: they prevent the development of axillary buds. If we cut a terminal bud the axillary buds will grow. They also help in the development of roots.

Nevertheless, a moderate dose can cause malformations.

A heavy dose arrests the appearance of buds.

A very heavy dose will kill a plant.

1.8.4 Other plant hormones

Recent investigations have enabled us to highlight the effect of gibberellins that play an important role in:

— cell multiplication,

— intercalary elongation of stems,

— growth of leaves, and

— breaking of dormancy (*see Chapter 6,* §*1.1.2.2b*), but have no effect on roots.

2 Physiology of Plants Lacking Chlorophyll

In §1, we have studied nutrition in green plants, plants with chlorophyll. Other plants are unable to ensure synthesis of their organic foods on their own because of a lack of chlorophyll.
They have been classified by their mode of living.

2.1 Saprophytic plants

These plants live only at the expense of inert organic matter.
They decompose dead organic matter and, as a result, are often more useful than harmful to man.
In this group we have the following:

2.1.1 Saprophytic fungi

Capped fungi on manure heaps, vegetable mould, stumps of trees (cultivated mushrooms, cepe).
White mildew on bread in a humid atmosphere.
Green mildew (*Penicillium notatum*), from which penicillin is extracted, is found on sour (acid) fruits.
Yeast, present on the surface of fruits, transforms glucose into alcohol: as in the manufacture of wine, cider and beer.

2.1.2 Saprophytic bacteria

There are bacilli of putrefaction; they produce heat (thermcgenic layers).
Micrococcus of ammonification: transformation of organic nitrogen of the soil into ammonia salts (see course on Soil Chemistry)
Various ferments:

— transformation of wine into vinegar,

— coagulation of milk,

— retting of flax.

2.2 Parasitic plants

These plants feed on living animals or plants. They destroy living organic matter.
The host plant reacts by activating its cell multiplication around the affected part to resist the action of the parasite.

An excrescence is formed but, in most cases, the host is unable to protect itself.
Some of these are:

2.2.1 Parasitic phanerogams

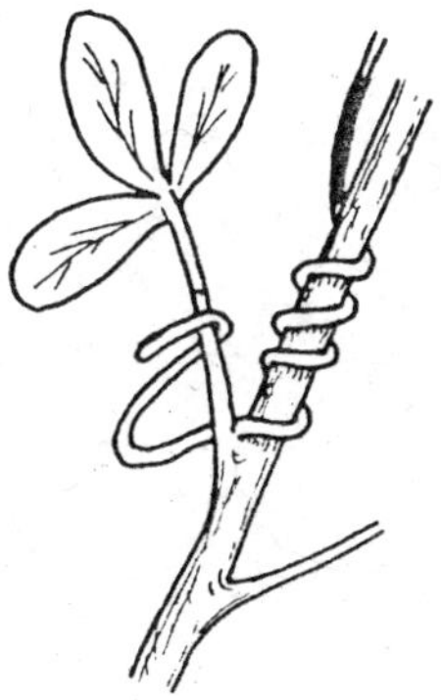

Fig. 5.12. Cuscuta on lucerne

a) An example is cuscuta (*Cuscuta*) on trefoil (*Trifolium*) and lucerne; it is a twining pink stem that inserts its haustoria into the host plant.
It absorbs both saps (raw and elaborated).
This is *total parasitism.*
The host plant dies.
b) Another example is mistletoe (*Viscum album*) on the poplar (*Populus alba*), apple (*Malus communis*) and sometimes the oak (*Quercus pedunculata*).
It may be noted that mistletoe, since it is green, has chlorophyll but, nevertheless, it is a parasitic plant because it is incapable of absorbing its raw sap from the soil: it draws sap from the host through its long sucker roots.
This is *partial parasitism.*
The sticky fruits become attached to the legs of birds and are, in this way, transported from one tree to another. The seeds germinate and the sucker roots penetrate other affected branches.
The host plant suffers on account of this parasitism.

2.2.2 Parasitic fungi

Merely as a reminder, because they are dealt with in more detail in the course on parasitology, we mention: mildews, oidium, scab, smut, etc.

2.2.3 Parasitic bacteria

In addition to destroying organic matter these bacteria produce toxins: diphtheria, tetanus, typhoid, etc., apart from bacteria that are parasitic on plants.

2.3 Symbiotic plants

Symbiosis is a reciprocal relationship in which each plant benefits from and supports the other.

2.3.1 Case of lichens

Let us again refer to our example of wall parmelia.
We had stated that this was a combination of two different plants: a fungus (*Ascomycetes*) and a unicellular alga (Chlorophyceae).

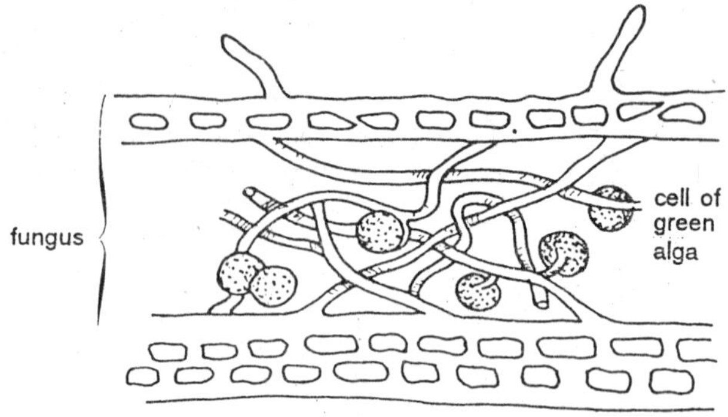

Fig. 5.13. The lichen

If we plant spores of fungi near cells of green algae we achieve *a synthesis of the lichen.*

We can also separately grow the alga and fungus of a lichen: we achieve *an analysis of the lichen.*

The fungus has:

— a mechanical role: fixation on a support,

— a physical role: protecting the alga from desiccation, absorbing water and mineral salts.

The alga has a chemical role: it ensures chlorophyll assimilation by its chlorophyll and, thereby, manufactures elaborated sap.

2.3.2 Case of mycorrhiza

These are fungi living on roots (as their name indicates) in the form of external filaments—beech (*Fagus sylvatica*), pine (*Pinus sylvestris*), oak (*Quercus pedunculata*)—or internal filaments (orchids) that also enable the host plants to absorb organic carbon directly, which would be impossible for them to do on their own.

2.3.3 Case of nodules in leguminous plants

Root nodules of leguminous plants contain bacteria known as "bacteria of leguminous plants", the *Rhizobium leguminosarum.*

These bacteria live in the soil as saprophytes until a leguminous plant is sown: pea (*Pisum sativum*), bean (*Phaseolus vulgaris*), broad bean (*Vicia faba*).

They then penetrate the cells of roots through the absorbent root hairs.

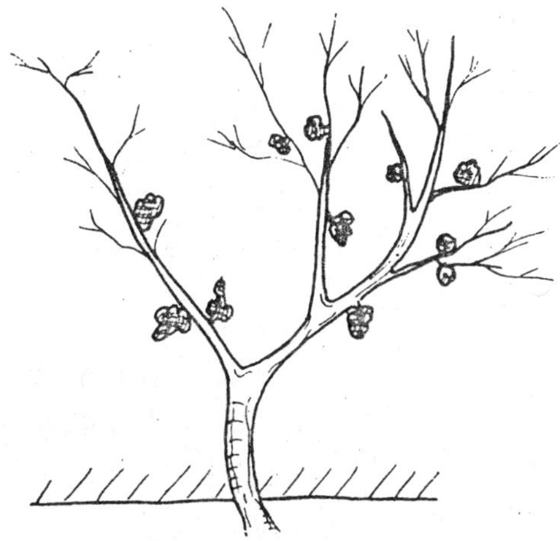

Fig. 5.14. Root nodules of leguminous plants

The host plant reacts to limit this "parasitism" and a hypertrophy of the cells results in nodules.
These bacteria are able to fix atmospheric nitrogen.
After this a symbiotic life begins.
The host plant provides carbohydrates.
The bacteria produce proteins with the help of nitrogen from the air, which are used by the plant.
This is therefore:

- an *attenuated parasitism*,
- tolerated by the host plant,
- essential for its normal growth,
- resulting in a state of equilibrium.

In a plantation of one hectare, it has been calculated that 200 kg of atmospheric nitrogen can be fixed in a year.[28]

28

a) When leguminous plants are grown, the requirement of fertilisers is limited.
b) When leguminous plants are to be harvested it is desirable to leave the roots in the soil.

3 Plant Movements

In spite of their relative immobility as compared to the possible locomotory movements of animals, we know that plants are able to move.
There are several types of movements due to different factors.

3.1 *Internal autonomous movements*

These movements take place without any external intervention.
Some of these are:
— Movements of the protoplasm in cells, visible under the microscope (*Chapter 3, §2.2.1.1*).
— Elongation of roots and stems seen in the experiment conducted with India ink marks (*Chapter 4, §1.1.3*).

3.2 *Movements due to external agents or considerations*

They occur on stimulation of an organ or a part of it.

3.2.1 Taxis or tactic movements

These refer to translatory movements oriented with respect to a stimulus.

3.2.1.1 Chemotaxis

a) Effect of oxygen
— Aerobic bacteria are attracted by oxygen.
— Anaerobic bacteria are repelled by it.
b) Effect of certain chemical compounds
Some male reproductive organs have cilia that help them to move when a secretion of malic acid is produced by the female organ:
— antherozoids of ferns
— zoo spores of algae.

3.2.1.2 Thermotaxis

Some bacteria move away from heat.

3.2.2 Tropisms

These involve movements that are directly related to the growth of vegetative organs of the plant.

3.2.2.1 Geotropism

The effect of gravity on growth:
a) of a root
Let us take a bean plant (*Phaseolus vulgaris*) that has just germinated and been pinned on a piece of cork. We place the young radicle with its tip facing upwards. After a day it bends and moves downwards.
The root undergoes the effect of gravity.
It has a *positive geotropism.*

Fig. 5.15. The geotropism of a root is positive. That of a stem is negative.

Note. The effect of gravity is less on secondary roots. This explains their oblique growth.

b) of a stem

In the same experiment, when the radicle was kept with the tip facing upwards we had the tigellum facing downwards.

When the root was inclined downwards at the end of the experiment the tigellum turned upwards.

The stem has a *negative geotropism.*[29]

29

Whatever the position of a seed in or on the soil, the radicle will always move downwards in the earth, and the tigellum will rise upwards.

"Orientations" of organs may subsequently modify these principles: for example, creeping plants.

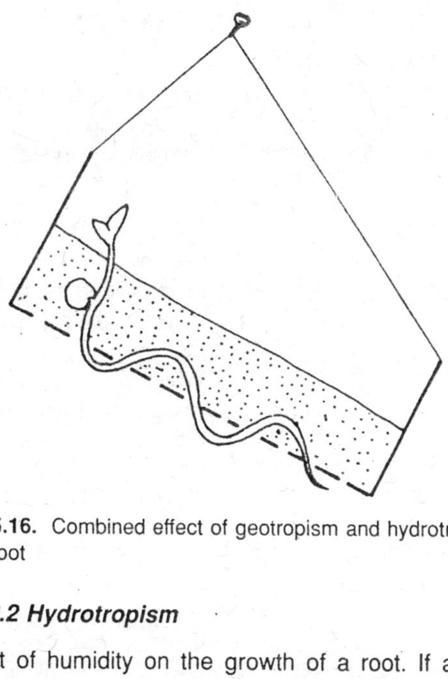

Fig. 5.16. Combined effect of geotropism and hydrotropism of a root

3.2.2.2 Hydrotropism

Effect of humidity on the growth of a root. If a root is exposed on one side to dry conditions and, on the other side, to humidity, we will soon observe it to turn

towards the humidity.
We say that *a root has a positive hydrotropism.*[30]
Experiment. (Combining the positive geotropism and hydrotropism of roots.)
We carry out the following experiment.
A bean seed (*Phaseolus vulgaris*) is placed in an inclined basket in which the bottom is replaced by a grill.
Due to positive geotropism the radicle will grow vertically downwards towards the bottom of the basket and will come out of the grill.
The humid area will then attract it due to positive hydrotropism, the radicle will turn towards the grill and enter the cotton wool. Geotropism will then cause the radicle to move downwards vertically ... and so on.

30
Near water, the roots of a tree always move towards it.
If, however, the humid zone is at the surface (watering) they rise towards it. This is seen in orchards cultivated by "mulching" or "irrigated" and in which the root system is at the surface. Here, we must be careful about agricultural practices (damage to roots).

3.2.2.3 Phototropism

The effect of light on the growth of a stem.
If we place a plant facing a window in an apartment we will find that, after a while, all its stems turn and are inclined towards the light.
The stem has a positive phototropism.[31]

31
In an apartment we must regularly turn potted plants. In a greenhouse it is desirable to have the light coming in from all sides.
This is one of the advantages of circular greenhouses.

3.2.2.4 Contact of the root with a solid

By a process similar to those of geotropism, phototropism or hydrotropism, when a root encounters an obstacle, it turns around it and continues to grow vertically.

3.2.2.5 How do we explain these tropisms?

We know that natural plant hormones (auxins) play a major role in the growth of stems and roots (§ *1.8.*).
It is the effect of these auxins that can explain such movements.
When parts of plants are exposed to light, water and the effect of gravity, their auxin concentration increases and, as a result, the affected sides grow at a slower rate than those that are not exposed.
As a result the organ faces one of the aforementioned sources and its initial growth is not restored until the auxin concentration on both sides is identical.

3.2.2.6 Elongation movement

During elongation of the root and the stem it is worthwhile to note that this growth occurs in the manner of a corkscrew.

This is what we call *circumnutation of the root and stem.*

3.2.3 Nastic movements

These apply to movements of leaves, flowers and stems that are independent of the phenomena of growth.

They are initiated by environmental factors that cause a release of a substance which acts on the motor cells due to a change in turgidity.

3.2.3.1 Effect of temperature

A flower blooms at a precise minimum temperature: crocus (*Crocus vernus*), water lily (*Nymphaea alba*).

The leaves of the robinia (*Robinia pseudo acacia*) fold up at about noon.

3.2.3.2 Effect of light

There are:

— day flowers that open in the morning and close in the evening: trefoil (*Trifolium repens*), lupin (*Lupinus polyphyllus*), convolvulus (*Convolvulus tricolor*).

— night flowers that open in the evening and close in the morning: the four o'clock flower (*Mirabilis jalapa*).

3.2.3.3 Effect of humidity

The involucre of the thistle (*Carduus nutans*) closes if the weather is extremely humid; on the other hand, it opens wide if the weather is dry.

This also applies to dehiscence in the fruits of the balsam (*Impatiens balsamina*).

3.2.3.4 Effect of orientation

The inflorescence of the sunflower (*Helianthus annuus*) follows the movement of the sun (whence its name).

Prickly lettuce (*Lactuca scariola*) has its leaves facing north and south, which enables it to benefit from a limited exposure to the rays of the sun at noon.

It has been called the "compass plant".

3.2.3.5 Effect of contact

a) Case of the sensitive plant (*Mimosa pudica*)

At a slight shock the composite leaves react by:

— folding their leaflets,
— drooping the leaflets along the central rachis,
— drooping the entire leaf towards the stem.

This is a rapid movement.

After a while the leaf is restored to its original position: the restoration is much slower.

b) Case of the dionaea or Venus fly trap (*Dionaea muscipula*)

At the slightest contact (a fly or a pencil), the leaf folds back in two, traps the fly and digests it.

c) Case of the urns at the ends of some leaves of nepenthes (the pitcher plant)

On contact, the lid that closes the pitcher folds back and traps the insect.

Chapter 6

Plant Reproduction

1

Sexual Reproduction of Phanerogams Anatomy and Physiology of their Reproductive Parts

1.1 Why does a plant flower?

A plant cannot flower unless it has the "aptitude" and, to have that, must fulfil three conditions:

a) It should have attained its *flowering maturity*. It may be recalled that Chapter 2, § 1.3 dealt with modes of plant growth: the begonia flowers in the year it germinates but the walnut does not flower for about ten years and, whatever the means adopted, it will never flower during the first or second year.

This flowering maturity depends on the plant species.

b) It must fulfil some *internal conditions* that are linked mainly with *problems of nutrition*.

c) It must undergo the influence of *external conditions* that are essential and specific for each species.

These conditions are governed by *thermal* and *light factors*.

We obviously cannot alter the age of flowering maturity.

On the other hand, knowing the exigencies of the internal and external conditions for flowering, we can find many horticultural applications.

1.1.1 Nutritional factors in flowering

The relationship between carbonaceous and nitrogenous food for a plant is most important. Good carbonaceous food alone will not help in flowering and, consequently, nitrogenous food should not be restricted.[1]

1.1.2 Thermal factors

These are conditions of temperature that occur at a

1

Any excess of nitrogenous fertiliser improves growth and adversely affects the flowering of a plant.

Therefore:

a) To obtain leafy vegetables we provide extra nitrogenous fertiliser.

b) We increase carbonaceous food for plants and protect their foliage by exposing it to the most intense illumination and enriching the environment with carbon dioxide.

These three means hasten chlorophyll assimilation (see Chapter 5, § 1.3).

It may be noted that an excessive lack of equilibrium between these two components could have effects contrary to what we would expect, by stopping normal growth.

regular rate, and this is why they are known as *thermoperiodisms.*

1.1.2.1 Diurnal thermoperiodism

Most plants originate from regions where there is a variation between day and night temperatures.

We know, from experience, that a plant grows well if we ensure a lower temperature at night than in the daytime.

Furthermore, it is mainly the night temperature that has an effect on flowering. If it is too high, flowering is delayed.[2]

1.1.2.2 Seasonal thermoperiodism

a) For the same ecological reasons, most plants, in their regions of origin, are subject to periods of activity and periods of rest.

This rest is called *dormancy.*

Our indigenous plants have a period of dormancy in winter during which they are subjected to conditions of temperature that are specific for each of them: stoppage of growth (deciduous leaves), followed by a period of fresh growth in the spring.

In any case, the period of activity can only begin at the end of dormancy and it can be seen that the future growth of a plant depends on how well this period of activity begins.[3]

b) We are aware that plants do not begin to grow in the spring unless their dormancy period is over: this is the *end of dormancy.*

It may be natural or artificial.

Cold treatment is, obviously, one of the means that all researchers have considered. We may recall the Russian experiments by Lyssenko on wheat (*Triticum sativum*).

We know that there is an autumn wheat (a biennial plant) and spring wheat (an annual plant).

The former needs a cold period to "flower", while the latter does not need one.

Nevertheless, autumn wheat, having been subjected to a period of low temperatures and sown in the spring, will flower during the same year and will behave like spring wheat.

This is *vernalisation.*[4]

Each plant species has its vernalisation temperature.

It may be noted that:

— depending on the plant species, vernalisation may be effected on the embryos or the seedlings,

2

Plants cultivated in greenhouses should not be subjected to a temperature that is constant both during the day and at night. About ten degrees is, generally, the desirable variation.

On the other hand, plants that have originated from equatorial regions where the day and night temperatures are the same are not affected by this difference.

3

Forcing is an operation that, primarily, involves subjecting plants to conditions closely resembling those which prevail at the beginning of their period of activity but with a considerable advancement or retardation in terms of time.

This, however, is only possible once the period of dormancy is over (end of winter) or if there is no such period (annual plants).

Except in the case of annual plants, this amounts to a hastened growth (we gain 1 to 3 or even 4 weeks) or a delayed growth (we delay production by several weeks) merely by cultural processes under cover that enable us:

— either to hasten, not the date, but the period of activity,

— or to delay the beginning of the dormancy period.

In forcing, growth is so much out of time that it is often required to begin during the dormancy period. Is this possible?

4

To the extent that plant species can adapt or horticultural techniques require, we can, by vernalisation:

— look for early maturing varieties by transformation of a biennial cycle to an annual one: lettuce (*Lactuca sativa*), for example:

— use autumn variety seeds, which have not proved successful, in the spring.

— the opposite operation may be applicable at a higher temperature; this is de-vernalisation with, if required, re-vernalisation.

c) Other methods than cold may, however, be adopted to "end dormancy": treatment of the plant in certain gaseous atmospheres, vapour baths, heat, desiccation (without, however, carrying this out to an extent that may kill the plant).[5]

Note on dormancy and forced rest

There should be no confusion with regard to the "dormancy period" due to physiological causes and "forced rest" due to exceptional external conditions that may be considered accidents of growth,[6] such as very dry weather or extreme cold.

Note on seasonal thermoperiodism

It is clear that the foregoing is not applicable to plants of equatorial origin, which grow throughout the year.

1.1.3 Light factors

The reaction of a plant to the relative duration of daylight and darkness in a period of 24 hours is called *photoperiodism.*

This periodicity determines the appearance of flowers. The natural requirements of plants with regard to the "total light" that they need every day in order to flower partly depend on their regions of origin.

Equatorial plants need as much light as darkness because in their regions day and night are of equal duration.

1.1.3.1 Classification of plants

a) Some plants, in order to flower, need a period of more than 12 hours of darkness per day: these are *short-day plants.*

b) Others need daylight for more than 12 hours: these are *long-day plants.*

c) Finally, others are indifferent and do not react to long or short days: rose (*Rosa canina*), carnation (*Dianthus caryophyllus*).

This explains why:

— short-day plants flower naturally in the spring and the autumn: chrysanthemum (*Chrysanthemum indicum*), dahlia (*Dahlia hortensis*), apple (*Malus communis*).

— long-day plants flower naturally in the summer: nasturtium (*Tropaeolum majus*), calceolaria (*Calceolaria hybrida*), pyrethrum (*Pyrethrum caucasicum*),

5

These are very important and form a basis for many forcings. For example:

— heat treatment of flower bulbs: several weeks at a specific temperature depending on the desired date of flowering: "prepared bulbs" of tulips (*Tulipa*) to flower at Christmas,

— treatment of rose plants (*Rosa*) with ethylene or natural gas,

— treatment of lilacs (*Syringa vulgaris*):

by cold: 1 to 2 months at 5–6°C,

by desiccation; several weeks at the end of summer,

by ether vapours, 48 hours,

by hydrocyanic acid vapours,

by vapour baths: 12 hours at 40°C.

— rhododendron (*Rhododendron indicum*) with hydrogen carbide, etc.

— the lily of the valley (*Convallaria majalis*) with hydrocyanic acid, etc.

The horticulturist has only to force the plant prepared in this manner.

6

In some cases these accidents of growth may be induced, especially to artificially delay the start of growth in the spring. The dormancy period is over but we delay the start of plant growth.

Horticulturists thus delay the date of flowering of the lilac (*Syringa vulgaris*), the lily (*Lilium candidum*) and lily of the valley (*Convallaria majalis*) by cold treatment.

Arboriculturists, using hormones that produce the same results as cold, can delay the start of growth in fruit trees by 15 days in the spring with an autumn spray. This results in less risk during spring frosts. Market gardeners prevent the initial formation of tubers on potato plants (*Solanum tuberosum*) with an "antigerm treatment".

lily (*Lilium candidum*), tuberous begonia (*Begonia erecta*), gloxinia (*Gloxinia speciosa type erecta*).[7]

1.1.3.2 Effect of photoperiodism

From experience, we know that a short-day plant subjected to a very short period of darkness as a result of artificial lighting will not flower.
The opposite applies to a long-day plant.[8]
Horticultural conclusions: Forcing of plants is dependent on their growth cycles, but a horticulturist can considerably influence:

- thermoperiodism (colder temperature at night than during the day, vernalisation, ending dormancy artificially),
- photoperiodism (short or long day).

1.2 Anatomy of the reproductive organs

1.2.1 Stamens

1.2.1.1 Filament

The filament has a petiolar structure. This is normal, every floral part being a modified leaf.

1.2.1.2 Anther

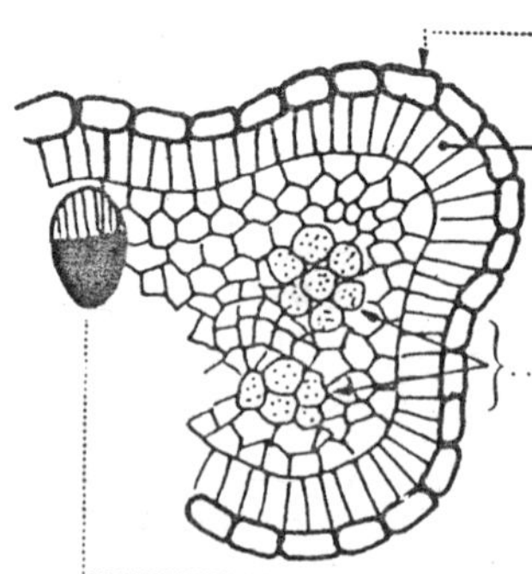

Fig. 6.1. Transverse section of an anther

In a transverse section of the anther of a wallflower (*Cherianthus chejri*), we observe a bilateral symmetry:

- Two swellings, one on each side: the *pollen chambers.*

7
This is one reason why some leafy vegetables should be planted on specific dates to prevent a premature and undesirable "run to seed".

8
We understand what the horticulturist can achieve from this condition of flowering.
Let us take the example of a short-day plant that flowers in the autumn, such as the chrysanthemum (*Chrysanthemum indicum*).

- If we expose it to light, its flowering will be delayed and may even be retarded till the spring.
- If we, at a very early stage, subject it to an extended period of darkness every day, its flowering will be hastened and may even occur in the summer rather than in the autumn (16 hours of darkness after the end of June = flowering advanced by 5 weeks).

Note: We must be logical. If we expose a plant to 13 hours of illumination per day, which for us is a long day, and if this plant needs 15, we will be subjecting it to a short-day treatment.
This horticultural practice is, therefore, useful provided we have a perfect biological knowledge of the plant.
Conclusive trials have already been obtained with geraniums (*Pelargonium zonale*), hydrangea (*Hydrangea hortensis*), tulips (*Tulipa*).

— Each of these chambers has two cavities: the *pollen sacs*.
— Within these sacs there is an accumulation of fertile cells, each of which divides into four cells. Each daughter cell, therefore, produces four *spores*, each of which has *n* chromosomes (by reductional mitosis or meiosis).

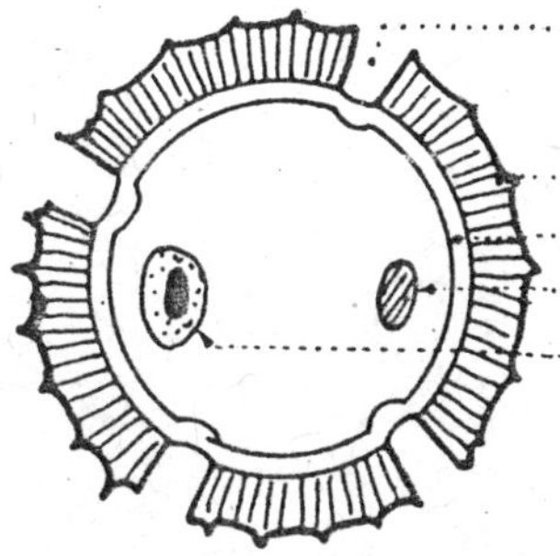

Fig. 6.2. Section of a pollen grain

— Each spore or "pollen grain" is composed of two membranes: exine and intine. The *exine* is external, resistant and cutinised. It is perforated in places, revealing "germ pores". The *intine* is internal, extensible and of cellulose, containing a *protoplasm* with two *nuclei* inside it, a *vegetative nucleus* and a *reproductive nucleus* or generative nucleus.
— An *epidermis* envelops the entire anther and the pollen sacs are surrounded on the outside by a "*mechanical layer*".
— In the centre, and extending from the filament, we observe a *vascular bundle*.

1.2.1.3 Dehiscence of the anther

When an anther is mature, the two mechanical layers are torn; opening the pollen sacs, they release pollen grains either from the top or from the sides.

1.2.2 Pistil

A pistil is formed from one or several carpels.
A carpel is a reduced leaf with a blade that bears ovules on its sides and is folded to enclose these ovules.

1.2.2.1 Composition of an ovule

An ovule is composed of two envelopes:
- an external envelope: *the primine* or outer integument,
- an internal envelope: *the secondine* or inner integument.

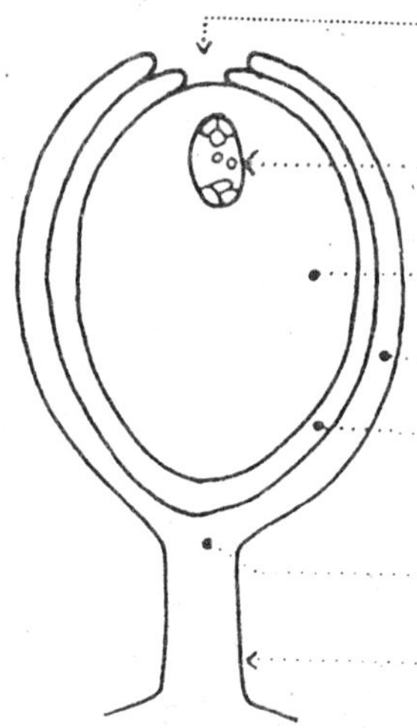

Fig. 6.3. An ovule

The primine and secondine are not joined at their top, there is a small opening: *this is the micropyle*, through which fertilisation will take place.
- The envelopes enclose a large mass, *the nucellus*, which soon produces a cell that divides into four cells with *n* chromosomes (by reductional mitosis or meiosis).

A single cell of the four survives and produces the macrospore in which the nucleus divides into eight parts contained in the embryo sac.
For fertilisation we are concerned with two parts: the oosphere and the cell sac.
- The ovule has a stalk: *the funiculus.*

The end of this funiculus, which remains as a scar on the seed, is called the *hilum*.
- The funiculus is crossed by a vascular strand that ramifies after the hilum at a point called *the chalaza.*

1.2.2.2 Position of the ovule

Depending on the respective positions of the hilum, the chalaza and the micropyle, we observe:
- erect or orthotropous ovules in rare cases: garden

sorrel (*Rumex acetosa*), sycamore (*Platanus occidentalis*);

— drooping or anatropous ovules: strawberry (*Fragaria vesca*), apple tree (*Malus communis*), potato (*Solanum tuberosum*);

— intermediate or campylotropous ovules: cabbage (*Brassica oleracea*), wallflower (*Cherianthus cheiri*), beet (*Beta vulgaris*), bean (*Phaseolus vulgaris*).

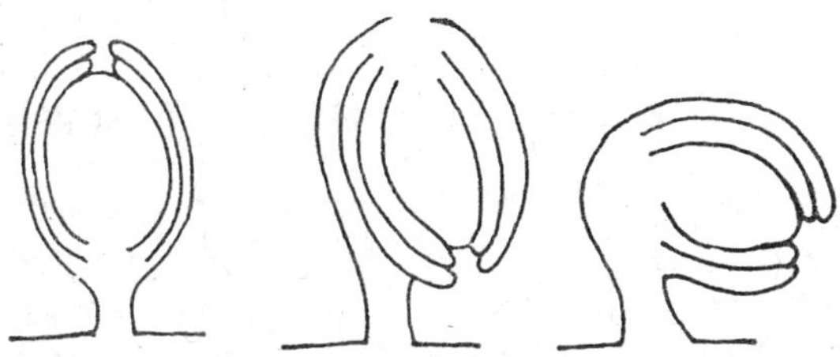

Fig. 6.4. Positions of ovules

1.2.3 Blooming of flowers

a) Flower opening depends on the species: it may take from a few minutes to two days.
b) The duration of blooming also varies: from a few hours to a month.[9]
c) Low temperatures delay blooming.[10]

9 This is of extreme importance when the flowering is required to be ornamental.
A plant that flowers for a very short period is of no ornamental value in flower beds.

10 Cold treatment may be a method of preservation in flower production (cut flowers) and in the florist trade by helping
— to extend the period of sale,
— in transportation.

1.3 Pollination

This involves transporting pollen grains, after dehiscence, from the anthers to the stigma of the ovary.

1.3.1 Modes of pollination

There are two modes of pollination.

1.3.1.1 Direct or self-pollination

In this the pollen of a flower arrives on the stigma of the same flower.
This occurs:
a) When the flowers are cleistogamous, that is, they do not open: violet (*Viola cornuta*), begonia (*Begonia gracilis*) to some extent.
b) When male and female organs mature at the same time: bean (*Phaseolus vulgaris*), pea (*Pisum sativum*).

It takes place:
a) Due to air movement: pea (*Pisum sativum*), bean (*Phaseolus vulgaris*).
b) By actual movement of the organs:—European barberry (*Berberis vulgaris*), in which the stamens drop on to the stigma.

1.3.1.2 Indirect or cross-fertilisation

This is the most frequent case, in which the pollen of a flower arrives on the stigma of another flower.
It occurs:
a) In all plants with unisexual flowers: in monoecious plants—oak (*Quercus pedunculata*), marrow (*Cucurbita pepo*)—and in dioecious plants—willow (*Salix babylonica*), poplar (*Populus nigra*).
b) In plants with bisexual flowers when:

— The male and female organs do not mature at the same time: usual case of *dichogamy*:
protandry if the stamens mature before the ovary (several composites and umbels),
protogyny, the opposite case (several graminaceous plants).

— The male and female organs do not have the same morphology: *heterostyly* :
long-styled flower, stamens lower than the stigma as in the potato (*Solanum tuberosum*),
short-styled flower, the opposite case, as in certain flowers of the primrose (*Primula hortensis*).

— The organs reveal cases of sterility:
intersterile plants, two varieties that cannot fertilise each other, and, obviously,
self-sterile plants, variety of pear cultivar Comice incapable of self-fertilisation: a secretion of the style prevents growth of the pollen tube.[11]

11
Selection and distribution of varieties is of major importance in an orchard when we must take into account:
- the dates of early or late flowering between varieties;
- the incompatibilities between male and female elements of the same variety, or between varieties.

1.3.2 Different types of pollination

1.3.2.1 Pollination due to gravity: direct pollination

The stamens are higher than the stigma.
On dehiscence from the anthers, pollen grains fall directly on the stigma.

1.3.2.2 Anemophilous pollination (by wind)

Oak (*Quercus pedunculata*), willow (*Salix babylonica*), hazel (*Corylus avellana*), pine (*Pinus sylvestris*).

Plants adapt to this manner of pollination when pollination is "somewhat infrequent" by:
a pollen that is easily carried by the wind (light pollen or tiny air balloons),
an enormous quantity of released pollen: "sulphur rain", a stigma that is highly receptive to this pollen.

1.3.2.3 Zoophilous pollination (by animals)

a) Entomophilous pollination by insects
Insects, slightly attracted by the colours of flowers but mainly by the sweet nectar formed by them, collect pollen and take it from one flower to another, the flower often adapting itself to receive it.[12]
Mobile stamens: vanilla (*Vanilla fragrans*), meadow sage (*Salvia pratensis*).
Hairs holding the insect till fertilisation: aristolochia (*Aristolochia durior*).
b) Ornithophilous pollination by birds
c) Zoidiophilous pollination by small animals: bats.

12
We note the advantage, for an arboriculturist, in having bees near his orchard.
On the other hand, we should be aware of the likelihood of unpleasant surprises: beds of pumpkin (*Curcurbita maxima*) or gourd (*Curcurbita pepo*) beside beds of melon (*Cucumis melo*).

1.3.2.4 Hydrophilous pollination

The same adaptations to water are involved as in plants that are subject to anemophilous pollination.

1.3.2.5 Pollination by man

This can be, unintentional but most frequently is deliberate and is called artificial pollination, as in the case of female date palms (*Phoenix dactylifera*) pollinated by Arabs.
This is also done in vanilla (*Vanilla fragrans*) plantations in Reunion Island where there is no insect that can fertilise these plants naturally, as in the West Indies.[13]

13
Artificial pollination is the method used by horticulturists to obtain new varieties (*see Genetics*) and by greenhouse tomato growers.

1.4 Fertilisation

A pollen grain is carried to the stigma of a flower: this is pollination.
It is retained by the viscous papillae and absorbs water by osmosis. It swells and germinates.
This germination takes several days (most frequent cases) or even a year: pine (*Pinus sylvestris*).

1.4.1 Germination of a pollen grain

The grain changes over from a slow to an active state of life.

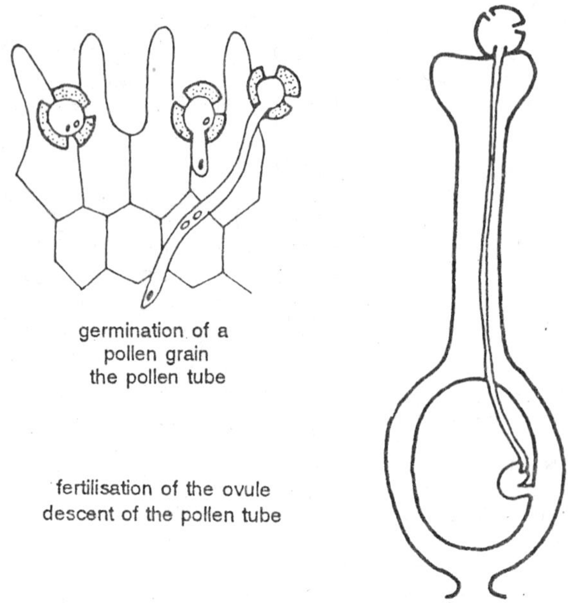

Fig. 6.5

Through a germ pore, a portion of the protoplasm emerges from its envelope and forms a pollen tube several centimetres long.

This tube enters the style and descends to the ovule, which it generally penetrates through the micropyle.

During its descent it secretes a hormone that helps in the fertilisation process.

A pollen tube comprises the two nuclei of the spore:

— the *vegetative nucleus*, which helps in growth of the tube and disappears before fertilisation,

— the *generative nucleus*, which divides, before fertilisation, into two parts, each containing *n* chromosomes (male gametes).

1.4.2 Actual fertilisation

One male gamete fuses with the oosphere or egg, which has *n* chromosomes, of the embryo sac and produces the *zygote* with 2*n* chromosomes.

The second male gamete fuses with the secondary nucleus of the embryo sac and produces the *endosperm nucleus.*[14]

14

Arboriculturists are all aware of the "washing away" of flowers, a physiological phenomenon of a serious type because it is, simply, a non-development of the flowers.

We are aware that this can be due to:

a) varietal incompatibility: self-sterility as well as external phenomena that hamper fertilisation.

b) an excess or a shortage of nitrogen that may prevent flowering and also hamper actual fertilisation.

c) bad weather:

— rains affect pollination and "drown" pollen grains,

— cold slows down the circulation of sap and prevents fertilisation (when too severe, below 2°C) and freezes the reproductive organs at 0°C.

We must remember that a good flowering may not ensure abundant fruiting.

1.5 Fructification

Once fertilisation has taken place, we know that a flower fades and the ovary develops as a fruit containing seeds formed by the ovules.
The increase in size of the ovary varies depending on the botanical types of fruits.
From the horticultural point of view only the development of fleshy fruits is of technical interest.
Let us study the case of a pear:

1.5.1 We first examine the stage of budding when we can be sure that fertilisation has been effective.

1.5.2 A tree always tends to retain a certain physiological equilibrium taking into account several elements, including its own production depending upon its capacity.
This results in a *natural fruit drop*,[15] a "personal selection" of the fruit tree.

1.5.3 The next stage is that of maturation together with several physiological and chemical phenomena.

1.5.3.1 Physiological phenomena observed in a fruit

Its temperature rises.
Its respiration is faster, resulting in a much greater absorption of oxygen and a release of a larger amount of carbon dioxide; the release of water vapour is also increased.
These phenomena are rendered more rapid by an increase in temperature, by the presence of oxygen and by a reduction in relative humidity, three external material causes that the professional has to cope with.[16]

1.5.3.2 Chemical phenomena

Acids in fruits combine with bases in the sap and produce salts resulting in a reduction of acidity in fruits.
Starch is transformed into sugar and the fruit becomes sweet, richer in pectin and more juicy.
Alcohols are transformed into essential oils, resulting in a better "fragrance" in fruits and also in volatile products resulting in a release of ethylene gas.[17]

15

An arboriculturist should adopt all possible means to reduce fruit drop during June ... in cases where budding has not been very abundant.
He must, especially, be attentive during an excess of dry weather in respect of a lack of plant food that may upset the equilibrium of a tree that has reached the stage of fruiting.

16

To preserve fleshy fruits:
- We must reduce the storage temperature: cold treatment.
- We must control the ventilation to reduce gaseous exchanges, and providing atmospheric nitrogen by reducing the proportion of oxygen is effective.
- We must ensure a certain amount of humidity to prevent an excessive loss of weight.

17

We can verify if a fruit is ripe with an iodised solution that turns blue in the presence of starch contained in unripe fruit.
For purposes of preservation, the phenomena of maturation being hastened by volatile gases (especially ethylene), such products must be removed from storage areas.
Since an increase in the percentage of carbon dioxide slows down ripening we can use the polyethylene bag technique; the bag is permeable to carbon dioxide, causing a delay in ripening but impermeable to water, resulting in a limited weight loss.

1.5.4 The fruit reaches a stage of maturity.[18]

1.5.5 At the point of insertion of the peduncle on the branch there is a formation of a point of abscission similar to the process in leaf fall: *this is called fruit drop.*[19]

Note on seedless fruits

We know that the movement of the pollen tube in the style is accompanied by secretion of a hormone from the tube that assists in fertilisation.[20]

18

Fruits known as summer and winter fruits:

- consumed in the same area are picked when ripe,
- when transported to a distance are picked before they ripen.

Winter fruits are collected before the frosts and ripen during preservation.

19

The use of hormones delays this abscission and helps to prevent premature fruit drop before the harvest (wind. . .)

20

Synthetic hormone sprays may, in some fruit species, initiate the process of fruiting but without fertilisation.

In such cases we obtain fruits without seeds, known as parthenocarpic fruits.

Experiments have been conducted with tomatoes (*Lycopersicum esculentum*) and with oranges (*Citrus sinensis*).

1.6 Seed

1.6.1 Formation of a seed

1.6.1.1 Endosperm

This is a reserve material formed by successive divisions of the endosperm nucleus till the sac is completely filled.

1.6.1.2 The embryo

The zygote divides in two:

- the lower portion produces the *embryo,*
- the upper portion provides the *suspensor* that forces the embryo into the endosperm.

1.6.2 Seed dispersal

This happens in many ways.

1.6.2.1 By wind

The seed or the fruit containing it undergoes an adaptation: samaras, hairs . . .

1.6.2.2 By animals

Hooked fruit—achene of the avens (*Geum chiloense*); ingestion of fruits and intact passage of seeds through the intestine—mulberry (*Morus alba*); fruit that sticks to the legs of birds—mistletoe (*Viscum album*).

1.6.2.3 By bursting

Balsam (*Impatiens balsamina*), broom (*Cytisus scoparius*).

1.6.2.4 By water

Seeds of the date palm (*Phoenix dactylifera*) are carried by the sea to some atolls in the Pacific.

1.6.2.5 By man

Aircraft, ships and other means of transportation.

1.7 Germination

1.7.1 Maturation of the seed

a) A fruit may ripen at the same time as the seed within it, as in wheat (*Triticum sativum*) or the bean (*Phaseolus vulgaris*).
On the other hand, the seed may not be ripe though the fruit is, and the variation may be as much as a year or even two, as in the peach (*Prunus persica*).[21]
b) Depending on their origin, seeds require a "quantum of heat" to ripen.[22]
c) Before they germinate, seeds are dormant. They respire very slowly, absorbing oxygen, and releasing carbon dioxide as well as water vapour.
They live at the expense of their reserve materials.
They also lose weight.
These phenomena are rendered more intense by a rapid change of air, an increase in temperature and a higher humidity.[23]

1.7.2 Suitable conditions for germination of a seed

1.7.2.1 It is clear that an embryo must be well formed and living, as it is to produce a seedling.[24]

1.7.2.2 The reserve materials should not have been tainted or allowed to deteriorate.

This is particularly difficult to ensure in the case of oil seeds that are prone to become rancid.

1.7.3 Modes of germination

Let us germinate three different seeds:
— one of the pea (*Pisum sativum*),
— one of the castor plant (*Ricinus communis*),
— one of the bean (*Phaseolus vulgaris*).

21
We may sow grains of wheat (*Triticum sativum*) or beans (*Phaseolus vulgaris*) immediately after a harvest. If it is technically possible, they will germinate.
In the case of seeds that do not ripen at the same time as the fruit, we have to wait till they reach the stage of maturity (peach: *Prunus persica*).
This period is known as the dormancy period of seeds.
We therefore adopt layering (*see the course material on plant propagation*).

22
This creates problems when planting varieties of maize (*Zea mays*) from the south of France in the north.

23
Seeds that are to be consumed are dried, stored in silos and even packed in insulating materials.
Seeds that are meant to be planted are to be moderately dried. Asphyxiation must be avoided.

24
The duration of life of an embryo is not unlimited whatever the methods adopted for storage.
This is what we call the germinative duration that may be from a few weeks to several years.
The horticulturist who is unsure of the germinative power of his seeds would find it worthwhile to conduct a germination trial (*see the course on plant propagation*).

1.7.3.1 Case of the castor oil seed (Ricinus communis)

The seed swells,
The integument is torn,
The radicle penetrates the soil.

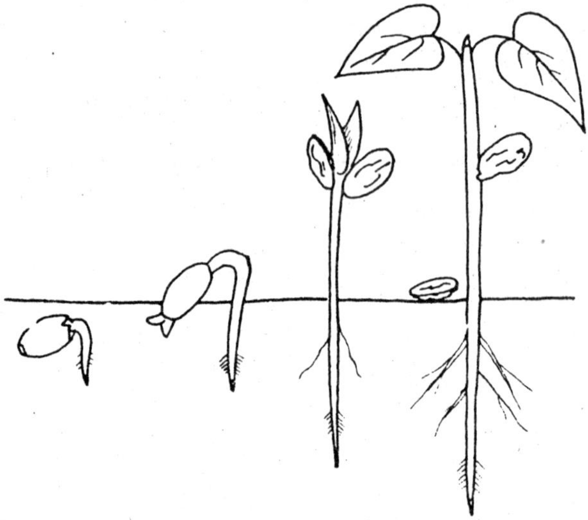

Fig. 6.6. Epigeal germination of a bean seed

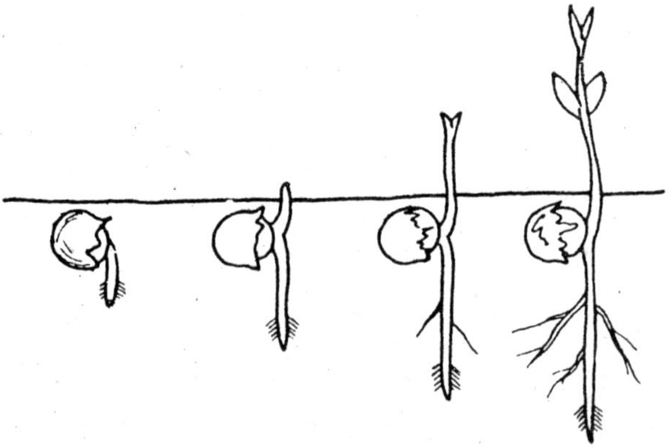

Fig. 6.7. Hypogeal germination of a pea

The seed is raised above the ground due to an elongation of the tigellum.
The cotyledons turn green and form the first two leaves.

1.7.3.2 Case of the bean (Phaseolus vulgaris)

The seed swells.
The integument is torn.
The tigellum raises the cotyledons, which separate.
The plumule appears and forms the first two leaves.
The cotyledons dry up and fall off, as their reserves have been exhausted.
In these two cases the only difference lies in the development of the cotyledons; because the seed was always raised above the ground, the hypocotyl below the cotyledons has elongated.
(By hypocotyl we refer to the portion of the tigellum between the radicle and the cotyledons.)
This was an epigeal germination.

1.7.3.3 Case of the pea (Pisum sativum)

The seed swells,
The integument is torn.
The radicle penetrates the soil but the seed is not raised.
The cotyledons remain in the soil.
The hypocotyl is not elongated.
This was a hypogeal germination.

1.7.4 Physiology of germination

1.7.4.1 To germinate, a seed requires the fulfilment of a certain number of external conditions.[25]

It needs:
a) Water, which plays:
— a mechanical role in the tearing of the integuments;
— a chemical role by dissolving reserve substances which the plant needs till it can cater for its own nutrition.
b) Heat: optimum 20°C.
c) Oxygen, because respiration becomes intense.

1.7.4.2 During germination

A seed respires considerably and releases heat.
It also transpires, resulting in a weight loss.
It digests its reserve materials with the help of enzymes that transform carbohydrates to glucose and aleurone to amino acids and saponify fats (*see Chapter 5*, § 1.6).
After this the plant becomes independent. It absorbs its mineral materials and water and ensures its chlorophyll assimilation.

25
After the seeds are planted they should always be watered unless we are depending on rain.
Seeds do not germinate at low temperatures.
An aerated soil is necessary for the existence of the embryo.

It may be noted that the processes are the same when planting bulbs, rhizomes and tubers. It is only the area of storage of reserves that is different.

1.8 Reproduction in gymnosperm

Gymnosperms are seed plants but their ovules are naked and there are no ovaries.
Let us revert to one of our earlier examples, the Scotch pine (*Pinus sylvestris*).
It is a monoecious plant.
On the same tree we observe reproductive structures called "cones" that are of different shapes and in different positions.
Cones are formed by scaly imbricated leaves.

1.8.1 Male cones

Erect, they are yellowish sporiferous spikes.
Each leaf of the cone harbours a stamen reduced to its anther.

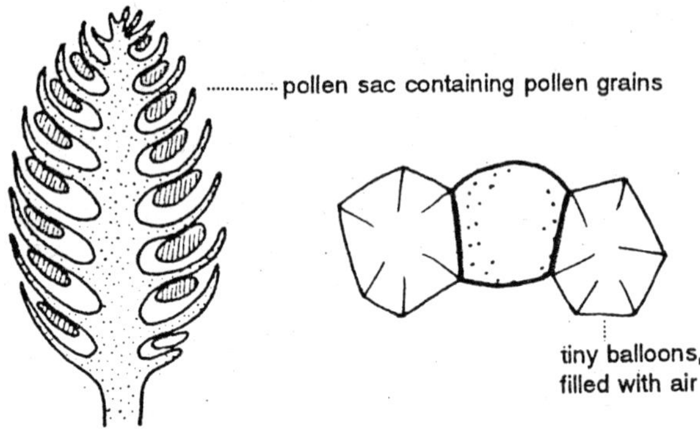

Fig. 6.8. Male pine cone and pollen grain

Each anther contains 64 *microspores.*
Every microspore is a pollen grain with four cells, of which only one will become an *antheridium.*
Each pollen grain is provided with tiny balloons filled with air (anemophilous pollination) and is dispersed at maturity.

1.8.2 Female cones

Erect, violet tinted, they are composed of "female flowers" formed from two scale leaves:

— one a bract scale, sterile scale leaf,

— one a carpellary scale, fertile scale leaf.

Each carpellary scale bears two megasporangia. Each megasporangium is an ovule in which the embryo sac contains *four eggs*.

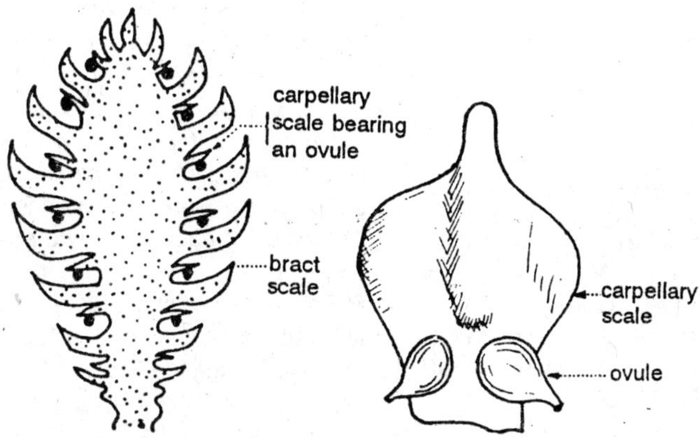

Fig. 6.9. Female cone and "female flower" of the pine

1.8.3 Fertilisation

Pollination and germination of a pollen grain, say, during the spring of 1987 would have resulted in fertilisation only in the spring of 1988, that is, in the second year.

An antheridium has two male gametes, of which only one will fertilise an egg to form a zygote.

A zygote produces four embryos. An ovule will, therefore, produce 16 embryos but only one will develop with many cotyledons within a winged seed.

After fertilisation the seed ripens in the cone and is then released due to a separation of the scales, but the empty cone remains attached to the tree.

On the pine, we therefore observe:

During the first year, female cones, erect, green and then brown and turned downwards after pollination, but only in the second year do we observe empty cones.

— A special feature of the fir (*Picea excelsia*): The scales of the cone separate individually. We never find dry cones below fir trees.

— Another special feature of the thuja (*Thuja occidentalis*) and the cypress (*Cupressus arizonica*): The duration of fertilisation is reduced from two years to one.

2 Asexual Reproduction of Phanerogams

■ Sexual reproduction involves fertilisation of female organs by male organs. Each of these has reproductive cells with n chromosomes that fuse to form a zygote with $2n$ chromosomes.[26]
A zygote, therefore, has chromosomes that originated from male organs and chromosomes from female organs. Consequently, it will have the hereditary characteristics of each of its two parents (*see Chapter 7, "Genetics"*).
■ Sexual multiplication is often a long process.[27]
■ What methods have been developed to ensure vegetative propagation?
We may adopt the following courses:
- Division of tufts: the stock is perennial and is broken up into smaller tufts. Example, perennial plants.
- Layering: a branch still joined to the rootstock is covered with earth: it develops roots and is then separated from the rootstock: rootstocks of the apple tree (wild apple) (*see fruit arboriculture*) and of the pear (wild quince) (*see fruit arboriculture*).
- Stem cuttings: an operation similar to layering but the branch is detached from the stock before roots have developed: geranium (*Pelargonium zonale*).
- Grafting: a portion of a plant detached from a stock is put in contact with another plant that has roots. There is a union.

One forms the underground portion and, sometimes, the stem. The other forms the aerial part and bears flowers and fruits. We graft cultivars of lilac (*Syringa vulgaris*), the apple tree (*Malus communis*), the pear tree (*Pyrus communis*) and the cherry tree (*Prunus avium*).
- *In vitro* cultures: these are being progressively developed and have now become industrial cultures.

Techniques of plant multiplication are described in the lessons on vegetative propagation.
We will deal only with the botanical aspect.

26
In certain cases, a horticulturist prefers to avoid these crossings (*see Genetics, Chapter 7*).
It was, therefore, necessary to find other methods of propagation without fertilisation.
Asexual reproduction has been developed and is known as vegetative propagation.

27
Vegetative propagation is very rapid.

2.1 Division of tuft

A perennial stock is separated and each small tuft branches out from the base with new stems emerging

from adventitious buds.
Adventitious roots also grow out.
This propagation is easy from the botanical point of view.

2.2 Layering

This involves the formation of adventitious roots, generally from buds on stems that are covered with soil.
The stem survives due to the sap it receives from the rootstock to which it is attached.
Some expedients help in the development of roots. These are:
- an incision that cuts the phloem strands: formation of an excrescence and, often, of roots;
- twisting of a branch, breaking the phloem elements;
- the use of hormones that help in the proliferation of certain tissues and especially of adventitious roots.

2.3 Stem cuttings

This is a more delicate operation because the new organs are to be formed exclusively from reserve material in the cutting, and the life of the stem is to be ensured after healing of the cut portion.
Healing by the formation of a suberified layer (as in the case of an injury) is rendered easier by a clean cut, avoiding damaged tissues.
The availability of reserve material in adequate quantity implies the use of well-formed branches and, above all, cuttings that are somewhat dry from branches that are well lignified.
The extended life of a cutting before it becomes "independent" necessitates a certain number of precautions in respect of "physiological economy", among which are:
- a reduction in leaf surface,
- a humid environment, except for some cuttings that are likely to rot.

The rapid growth of roots that occurs only after formation of an excrescence and adventitious roots from the central cylinder, like that of all root systems, may also be facilitated by the use of hormones.

2.4 Grafting

This is the most delicate method of propagation.
The essential requirement is to ensure a union.

The union is formed by the two generative rings that multiply rapidly till the end of this operation, which finally involves connecting, on one hand, the phloem of the stock and the scion and, on the other hand, the wood of the stock and the scion to ensure a continuity in the circulation of sap.
We understand why:

— If the affinity is not very good, the development of generative rings will be considerable, resulting in a bulk of healing growth.
— The cambium of the main generative rings involved in the union must coincide absolutely between the stock and the scion at least at one point: in grafting it is not the barks that have to be aligned but the cambiums.

There is a problem that is interesting to imagine: it is what the technicians call "the graft hybrid".
We have mentioned that the scion produces the aerial part with leaves, flowers and fruit (this, at least, is the object that grafters attempt to achieve).
Nevertheless, European vines grafted on American vines after an attack of phylloxera at the end of the 19th century did not produce quite the same grapes.
Moreover, in the Botanical Garden of Paris, grafting a yellow and a pink cystus produced a stock that had flowers of both colours.
From this should we assume that, in certain cases, hereditary characteristics are transmitted to the scion? No, this is what geneticists call a chimera (*see Chapter 7*).
In any case, the qualities of the stock influence the growth of a grafted plant in respect of:

— the comparative vigour of the tree,
— the extent of coloration of the fruit, etc.

Finally, it may be borne in mind that systematic relationships have nothing to do with the affinity of plants for grafting, which would appear to be paradoxical.
We may refer to the case of the pear tree, which is very close to that of the apple tree; the two can be grafted on each other only with great difficulty.
Nevertheless, the pear tree, "systematically" less related to the quince, can be successfully grafted on this plant species, which, in commercial growing, has become its rootstock.

2.5 *In vitro cultures*

These are modern methodologies that facilitate ultra-rapid multiplications in bulk to obtain a large number of

Fig. 6.10. Mountain lily being propagated *in vitro* (photo L.R.P.V.)

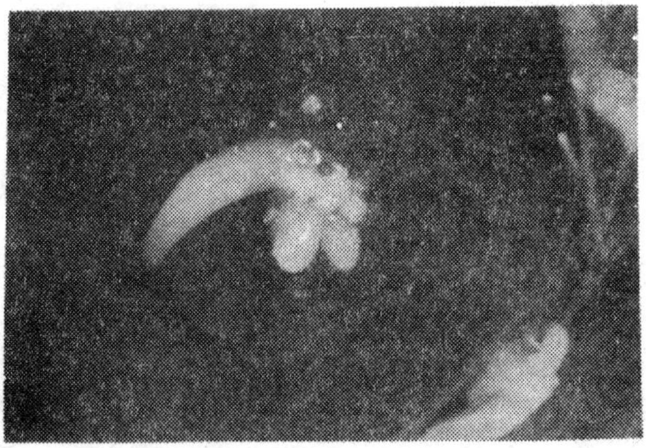

Fig. 6.11. Cluster of somatic embryos of the date palm (photo L.R.P.V.)

new varieties and/or plants that are resistant to virus infections (improvement of plants and regeneration). These cultures are in respect of organs, meristems and cells.

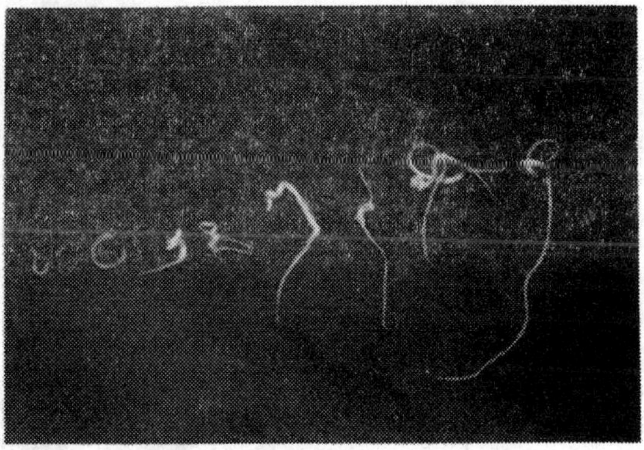

Fig. 6.12. Somatic embryos of the date palm: various stages of development (photo L.R.P.V.)

The basic principle, for example, on meristems involves:

— Preparing the culture medium (aqueous or solidified solution: gel),

— Sterilising this medium by using an autoclave (20 minutes at 120°), and subsequent cooling by passing it through a gas chamber or under ultraviolet rays.

Fig. 6.13. *In vitro* formation of roots of jojoba (photo L.R.P.V.)

Fig. 6.14. Replanting in sterile conditions under a canopy with a luminous flux (photo L.R.P.V.)

Fig. 6.15. Culture hall (photo L.R.P.V.)

Fig. 6.16. Sterilisation of culture mediums using an autoclave (photo L.R.P.V.)

— Sterilising the tissues and replanting them under a sterile canopy in tubes that are then kept in sterile conditions of illumination at 2500 lux and at a temperature of 20 to 25°C.
— Keeping the products obtained in culture halls in larger tubes or flasks to obtain an elongation and roots of 5 to 30 plants at a time.
— Placing everything in post *in vitro* culture with 3/4 of the container filled.
— Ending the operation by acclimation in *in vivo* culture.

3 Reproduction in Cryptogams

In this study we will deal with the same examples that we dealt with in morphological and anatomical studies of non-flowering plants.

3.1 Ferns

3.1.1 *Polypodium aureum*

Under the fronds we noticed brown spots. These are *sori*, which appear at the end of summer.

Each sorus is composed of a group of *sporangia*. Every sporangium has 16 fertile cells and each cell produces 4 spores.

The spores resemble pollen grains but their role is different.

In fact, a spore will germinate on soil and produce a small, green, heart-shaped blade, *the prothallus*, which has no stems, leaves or roots but, on its lower surface, will have male and female organs.

Antheridia (male organs) open out and release *antherozoids* with motile cilia.

These male gametes have only *n* chromosomes.

The *archegonia* (female organs) contain *eggs* with *n* chromosomes.

It may be noted that archegonia secrete malic acid said to attract antherozoids (*see Chapter* 5, § 3.2.1.1).

In damp weather the fusion of an antherozoid and an egg results in *a zygote* with 2*n* chromosomes.

This zygote with four cells produces the *embryo*.

One cell of the young fern forms the foot that feeds the embryo,

another is the origin of the root,

a third produces the stem,

and the fourth, the leaf.

3.1.2 Plants similar to ferns

a) The horsetails have the same multiplication process but the prothallus is unisexual.

b) The selaginellas (*Selaginella cuspidata*), like the horsetails, have a unisexual prothallus but their reproductive organs are composed of sporiferous spikes in which the base is formed of sterile leaves and the top of fertile leaves that produce micro- and megasporangia.

3.2 Mosses

3.2.1 *Sphagnum cymbiifolium*: peat moss

Some stems bear *sporangia* at the end with a mass of *antheridia*, male organs, containing many *antherozoids* and bottle-shaped *archegonia* that attract antherozoids by saps and contain eggs.

After fertilisation, a zygote is formed that germinates and produces a *sporogonium*.

This sporogonium lives as a parasite at the top of the leafy stem and produces *spores*.

Each spore produces a new moss by germination.

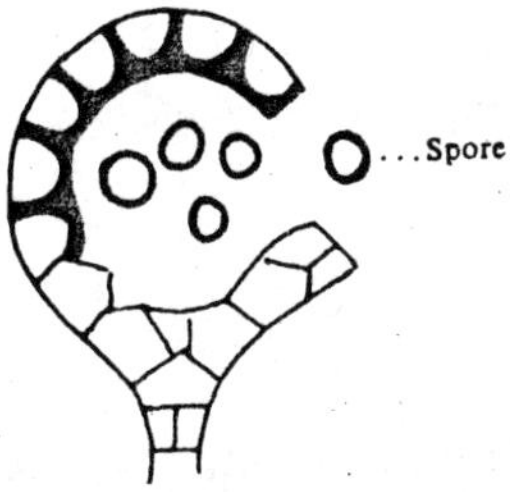

sporangium

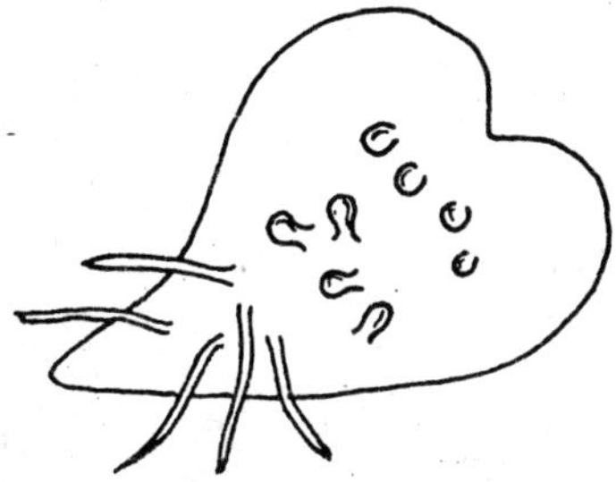

spore prothallus
with male and female organs

antherozoid

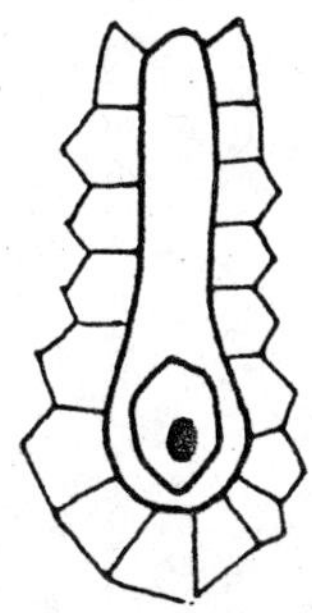

archegonium

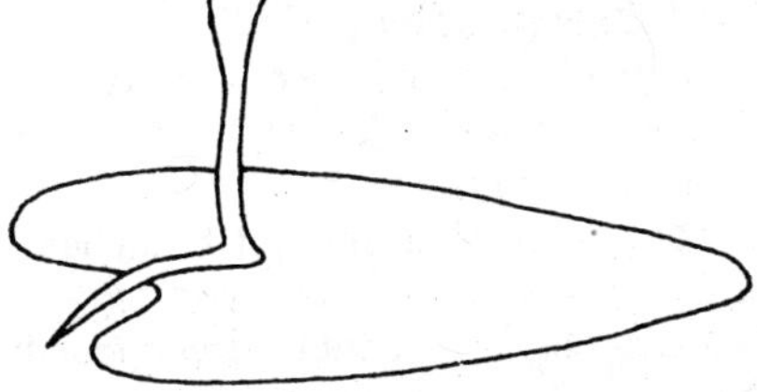

embryo

Fig. 6.17. Sexual reproduction of the polypodium

Male and female gametes have n chromosomes. The sporogonium and the zygote have $2n$ chromosomes.

3.2.2 Plants similar to mosses

Liverworts (hepatics) have an identical process of multiplication.

3.3 Fungi

3.3.1 The edible mushroom or Agaric

As in all capped mushrooms, we are aware that the aerial part of the plant is the reproductive organ.

The lamellae under the "cap" have *basidia*, the fruits of this type of mushroom (which is, therefore, called *Basidiomycetes*).

Each basidium contains two spores: the basidiospores, with two nuclei which, at maturity, render the lamellae brown.

Germination of basidiospores produces a new mycelium: the vegetative organ.

It may be noted that this fungus can be propagated vegetatively by cuttings in certain environmental conditions (*see growing of the edible mushroom*).

3.3.2 Scab on the apple tree (*Venturia inaequalis*)

This is, evidently, a type of multiplication in which the arboriculturist is greatly interested.

The economic importance of the parasite is, however, such that this aspect is dealt with in detail in the course on parasitology,

— in respect of treatment,
— as well as with regard to the development and, therefore, the reproduction of this fungus.

We recommend a study of that book.

3.4 Algae

3.4.1 Brown seaweed

We may come across two types of Fucus:

— the monoecious Fucus,
— the dioecious Fucus.

In both cases, the ends of branches produce fertile swellings containing male and female conceptacles.

Male conceptacles contain *antheridia* that produce *antherozoids*.

Fertilisation results in *a zygote*, from which a young alga originates (*see Fig. 6.18*).

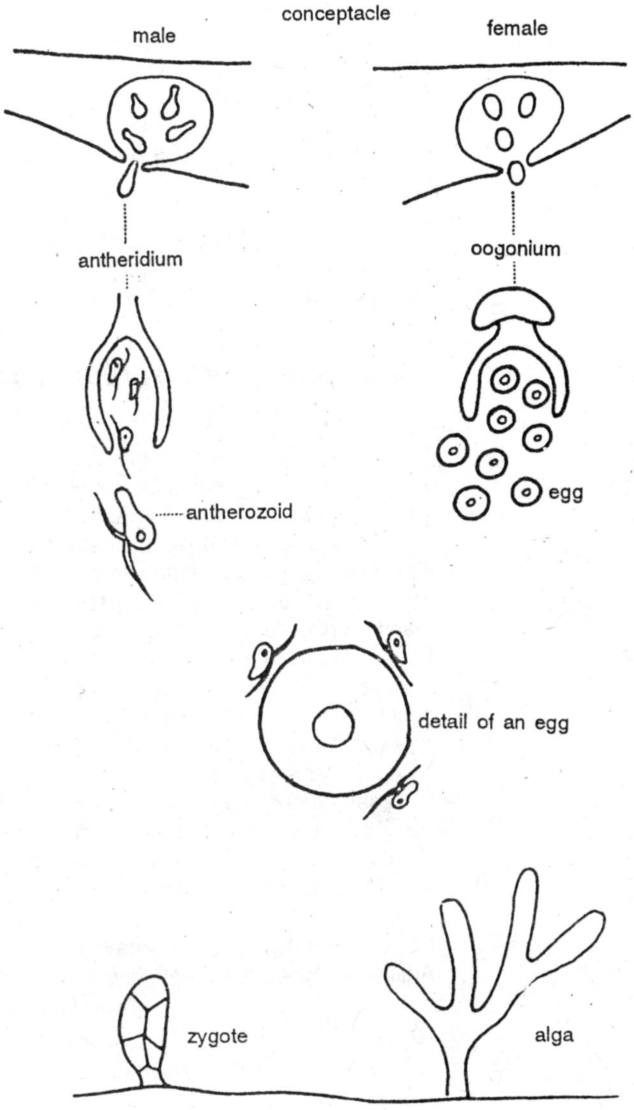

Fig. 6.18. Sexual reproduction of the Fucus

3.4.2 Microscopic algae

These may multiply by spores without fertilisation, as in protococcoidal forms: a green powder seen on the boles of trees.

3.5 Lichens

Lichens like wall parmelia are the result of an association of an alga and a fungus.
As such they can multiply:
— by *spores*, but in this case the fungus will multiply (fungus spore) because it very rarely has an alga near it.
— by *natural fragmentation*, by organs called *soridia* and comprising a network of filaments of the fungus and green cells of the alga.
This method helps the Parmelia to invade walls.

3.6 Comparative reproduction of cryptogams

In ferns. The adult plant originates from a zygote develops in a sexual prothallus which itself arises from spores produced by an adult fern.
In mosses. The adult plant originates from a spore that develops in a sporogonium which itself results from a zygote produced by an adult moss: in other words, in a manner opposite to that of the fern, which is produced from a zygote.
In fungi. A thallus from the reproductive organs produces spores that reproduce this thallus.
In algae. Several modes of propagation are possible:
— either, as in the case of fungi: alga, reproductive organ, spore, alga;
— or by fertilisation: alga, sexual conceptacle, zygote, alga;
— or both at the same time.
In lichens.
Normally, by natural fragmentation.
Artificially, by spores (*see Chapter 5, § 3.1*).

Chapter 7

GENETICS

1
Plant Genetics

Plant genetics is the science that examines all aspects of heredity and in which work is based mainly on *observation and experimentation.*

Mendel, a botanist monk who lived during the 19th century, was the forerunner of plant genetics.

2
Heredity

Heredity involves a set of genetic characteristics transmitted by ancestors to descendants in the animal and plant kingdoms.

2.1 Hereditary characteristics

2.1.1 Location of hereditary characteristics

These characteristics are located on the chromosomes in the nucleus of the cell.

In fact, each chromosome carries genes and every gene supports a given hereditary characteristic.

On fertilisation of an ovule there is a fusion of the male reproductive cell containing n chromosomes with the female reproductive cell that has n chromosomes (*Chapter 6, § 1.4*).

Each male chromosome has its female homologue. The fusion of these sexual gametes results in a zygote with $2n$ chromosomes.
As a result, the pairs of chromosomes in zygotes transmit hereditary characteristics of the parents to their progeny.
Nevertheless, a faithful transmission is effected only to varying extents.

1
If a market gardener sows dwarf plant seeds of round peas he will harvest round seeds from dwarf plants.

2.1.1.1 Sometimes the characteristics of parents are transmitted wholly.[1]

2.1.1.2 Nevertheless, the characteristics of parent plants are not always visibly transmitted to their progeny.[2]

2
The arboriculturist is well aware of the fact that when, for example, seeds of the pear from a Williams cultivar are sown, the trees that grow from these seeds will, obviously, produce pear trees but the fruits will be very different from those produced by the original tree.

2.1.2 Latent and manifest hereditary characteristics

All hereditary characteristics do not, therefore, appear to be similar to those of the parents.

2.1.2.1 When we planted seeds of Williams pears, we obtained plants that were different from each other and also different from the parent plants, even if only in their fruits (shape, colour, taste ...).

Each individual plant thus obtained was *a phenotype*, implying an indication of external appearances, manifestations of apparent characteristics and even non-hereditary *characteristics*.

2.1.2.2 Nevertheless, each chromosome of this phenotype had all the characteristics of the parent plants, including those that are not apparent: *the latent characteristics*.

All these latent and manifest characteristics together constitute a veritable inheritance of an individual: this is *the genotype*.

2.1.3 Transmission of hereditary characteristics

We are, therefore, aware that an embryo receives n chromosomes from the male parent and another n chromosomes from the female parent.
Each male chromosome corresponds to a female chromosome.
Every gene in one of the male chromosomes corresponds to a gene of a female chromosome.

The characteristics as a whole are, however, transmitted randomly.

2.1.3.1 If the same characteristic is located on a homologous chromosome *the plant is homozygous.*[3]

3 Multiplication: of the pea, the violet, etc.

Descendants resemble their parents. We have a *pure line* that we can propagate by the sexual method without risk of a "surprise" among the descendants. The genotype is similar to the phenotype.

2.1.3.2 If different characteristics are located on a homologous chromosome, *the plant is heterozygous.*[4]

4 Multiplication: of the potato, fruit trees, the dahlia, the geranium, etc.

Descendants do not necessarily resemble their parents.
The only way to propagate these "individuals" while retaining their specificity is to adopt the vegetative method.
The phenotype resulting from sexual multiplication will cause the appearance of only some of the possibilities of the genotype.

2.2 Non-hereditary characteristics

These have nothing to do with hereditary characteristics. They are observed in one or several individuals and, most often, are a result of a change in environmental conditions.
For example, in a bed:

— a particular plant may grow more vigorously than others;

— in a part of a garden there may be a more luxurious patch of vegetation;

— the same plants from the same pure lines or the same vegetative propagations may produce different results in different parts of a garden.

This should not lead us to believe that we are dealing with a new type of plant but merely that the individual or individuals have benefited from particular conditions such as:

— a larger amount of fertiliser,

— a particular microclimate,

— different cultural conditions, etc.

These changes are ... unfortunately ... in no way hereditary!

3
Mendel's Laws

Mendel was a monk who lived in Moravia during the 19th century.
He was extremely interested in botany and conducted many experiments on the basis of which he developed laws on heredity that came to be accepted only during the following century.
He studied crossings between plants of pure races with one or several pairs of different characteristics.

3.1 Monohybridism

Let us first study the simplest case in which a single pair of characteristics is different. This is monohybridism.

3.1.1 Case where first-generation hybrids are similar to their parents

Let us cross a round pea with a wrinkled pea.
Let us indicate the parents by P, the successive generations by F, round peas by O and wrinkled peas by ●.

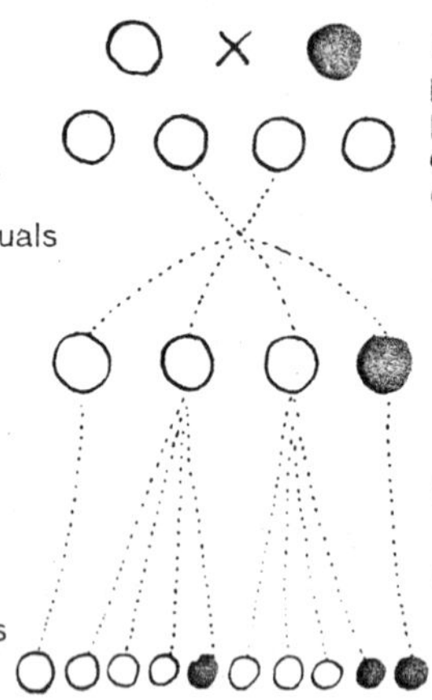

Fig. 7.1.

In F_3 what have actually appeared are:

— 1/4 of individuals of a pure race with a dominant characteristic that will always produce individuals of a pure race with a dominant characteristic.

— 1/4 of individuals of a pure race with a recessive characteristic that will always produce individuals of a pure race with a recessive characteristic.

— 2/4 of heterogeneous individuals whose descendants will be like those in F_2.

In other words, the ratio will be 1-2-1.

Note on the parents. The results obtained in this experiment are identical whatever the parent selected as father or mother.

3.1.2 Case where first-generation hybrids differ from their parents

Let us now study another case of monohybridism:
The *Mirabilis jalapa*, the four o'clock flower plant. We select a pure line with white flowers and a pure line with red flowers.

Let us pollinate flowers of the parent with white flowers PW with pollen from flowers of the parent with red flowers PR.

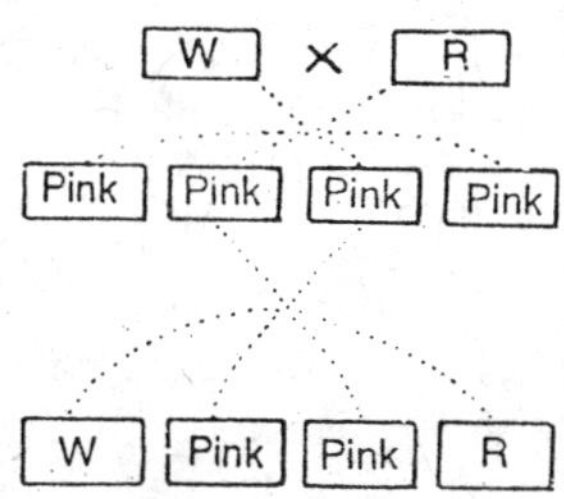

We obtain an F_1 flower — either a homogeneous phenotype, but different from the parents

Let us cross the individuals of F_1 with each other

We obtain: an F_2 — or a heterogeneous phenotype.

With
1/4 of W identical to the father
1/4 of R identical to the mother
2/4 of Pink identical to F_1

Fig. 7.2.

In F_3, the descendants of F_2 crossed by lines reveal:

— That individual specimens produced by the four o'clock plant with white and red flowers are of a pure race that is stable, and have the characteristics of one of the parents.

— That individual descendants of the four o'clock plant with pink flowers are heterogeneous and behave as do those of F_1, that is, they will produce:

1/4 with white flowers,
1/4 with red flowers,
1/2 with pink flowers.

The ratios are always the same as those in monohybridism, i.e., 1-2-1.

In the case of the *Mirabilis jalapa*, the characteristics in respect of white and red flowers are neither recessive nor dominant. They reveal an incomplete dominance.

Note on the parents. For example, in the pea, the results obtained are identical whatever the parent selected as father or mother.[5]

Comment. We always represent a dominant characteristic by a capital letter and a recessive characteristic by a small letter.

As a result, pure lines are written with two capital letters RR or two small letters *rr*.

3.2 Dihybridism

In a crossing it may happen that two pairs of characteristics are different. This is what is called dihybridism.

Mendel conducted his experiments with peas:

— one of selected parents had round yellow seeds: symbol RY.

— the other had wrinkled green seeds: symbol *ry*.

The capital letters R and Y represent dominant characteristics.

The small letters *r* and *y* represent recessive characteristics.

In the case under consideration, for $2n$ chromosomes the characteristics of the parents may be symbolised as follows: RRYY and *rryy*.

The gametes of each parent being RY and *ry*,

F_1 is symbolised by R-*r*-Y-*y*.

Only dominant characteristics are visible. The phenotype will, therefore, be RY and all the seeds obtained will be round and yellow, similar to those of the parent plants.

The possible gametes of F_1 will, therefore, be R-Y, R-*y*, *r*-Y, *r*-y.

Mathematically we will, in F_2, have 16 possible combinations.

	RY	R*y*	*r*Y	*ry*
RY	(1) RRYY	(2) RRY*y*	(3) R*r*YY	(4) R*r*Y*y*
R*y*	(5) RRY*y*	(6) RR*yy*	(7) R*r*Y*y*	(8) R*ryy*
*r*Y	(9) R*r*YY	(10) R*r*Y*y*	(11) *rr*YY	(12) *rr*Y*y*
ry	(13) R*r*Y*y*	(14) R*ryy*	(15) *rr*Y*y*	(16) *rryy*

(1-2-3, etc. are the case numbers)

We have, however, only nine different genotypes.

5

In the second case, distinction between individuals of a pure race and others is rendered easier in the second generation.

In the first case, we have to wait for another generation to be sure of the fixation of a variety.

Why?

Because hereditary characteristics are transmitted according to the laws of chance.

Because we have just observed that there are two types of characteristics: dominant and recessive.

When the former exist, even in a single case, they determine the phenotype.

If a pea has both "round" and "wrinkled" characteristics, only the "round" characteristic is visible.

When recessive characteristics exist they must do so in totality to produce a phenotype.

If a pea has a wrinkled characteristic in duplicate, the phenotype is wrinkled.

In the second example of the four o'clock plant, the characteristics are neither dominant nor recessive.

Case 1 repeated 1 time
Case 2 repeated 2 times
Case 3 repeated 2 times
Case 4 repeated 4 times
Case 6 repeated 1 time
Case 8 repeated 2 times
Case 11 repeated 1 time
Case 12 repeated 2 times
Case 16 repeated 1 time

Of these 16 possibilities, only 4 are of pure race and determine the finally fixed homozygous varieties.

1 bi-recessive *rryy*	phenotype of one parent
1 monorecessive *rr*YY	phenotypes different
1 mono-dominant RR*yy*	from the parents
1 bi-dominant RRYY	phenotype of one parent

The other 12 formulae determine non-fixed heterozygous varieties: these are hybrids.
Varieties of a phenotype different from the parents are pure, fixed and new.[6]
The independence of the two characteristics *vis-à-vis* each other may also be noted.

6 Here we observe a creation of new varieties by hybridisation that is useful for growing vegetables, fruits and ornamental plants.

3.3 Polyhybridism

This refers to situations where several pairs of characteristics are different.
It is clear that the number of possible combinations is in geometrical progression (64 for three pairs of factors) and this partly explains the difficulties encountered by our hybridisers.

3.4 Conclusions on Mendel's laws

3.4.1 Mendel's laws are applicable to all living organisms, both animal and plant, whatever the factor under consideration: size, colour, hairiness, etc.

3.4.2 If a factor is dominant, it is the only one that will appear in the first generation (F_1).

3.4.3 There is a segregation of characteristics in the second generation (F_2).

3.4.4 There is an independence of characteristics between each pair of them.

4 Hybridisation

4.1 Principles of hybridisation

4.1.1 Knowing Mendel's laws it is possible to carry out a theoretical study of the descendants of two specific parents.
Depending on his objective, the hybridiser will select:
the male parent P,
the female parent P,
knowing that *hybridisation* does not result in new characteristics (as in mutation) but enables new combinations of characteristics from those possessed by both parents.
We have to wait till generation F_2 to note the segregation because F_1 will be similar to the parents, whether male or female (except when the characteristics are neither dominant nor recessive, as in the case of the four o'clock plant).

4.1.2 Affinities

4.1.2.1 Between individuals of the same species.

There is, generally, no problem in such cases and fertilisation is possible.
This applies to cultivars of the apple tree, among others, Golden Delicious, Queen of Rennets (except in cases of intersterility: *see fruit arboriculture*).

4.1.2.2 Between individuals of different species, crossing is rarely possible. This impossibility may be due to the fact that hybridisation:

— is not followed by fertilisation,
— has not produced seeds,
— has resulted in sterile seeds,
— has produced seeds of non-viable seedlings.

There are exceptions: this applies to some species of orchids.

4.2 Techniques of hybridisation

These involve transporting the pollen of a "father plant" to the stigma of a "mother plant".
Two cases need to be considered:

4.2.1 Case of unisexual plants

The operation is facilitated because we have only to ensure protection for the "mother plant" to avoid any alien pollination and to transport the selected pollen to the stigma of the "mother flower".

4.2.2 Case of hermaphrodite plants

The operation is a more delicate one because, before pollinating with pollen from the "father plant", the stamens of flowers on the "mother plant" must be removed, an operation that is called castration or emasculation. The "mother plant" must then be protected against any external pollination and pollination with pollen from the "father" is to be ensured.
In both cases it is clear that:

— the male and female organs must be mature,
— after pollination the "mother plant" must be protected against any external pollination (gauze covers, cages ...).

4.3 Heterosis

Repeated propagation by self-fertilisation in each generation results in considerable consanguinity among the descendants.
The varieties obtained in this manner are very pure because we have increased the chances of appearance of a homozygous individual that could be a starting point of a pure line: this is inbreeding.
At the same time, the vigour of the next generation is reduced.
Nevertheless, if the hybridiser selects parents from pure lines when carrying out hybridisation, the first generation F_1 would appear to be much more vigorous than the parents: this is what is known as *hybrid vigour* or *heterosis*, but the second generation F_2 (and, especially, subsequent ones) will not retain these particular qualities.
Conclusion. The phenomenon of heterosis is manifest only in the F_1 generation.[7]

7

Hybrids of pure lines in F_1 produce very useful individuals in horticulture due to their vigour: flowers in larger numbers, of larger size and with a better colouring ...
This is observed in the zinnia (*Zinnia elegans*), sage (*Salvia splendens*), snapdragon (*Antirrhinum majus*), petunia (*Petunia grandiflora*), marigold (*Calendula officinalis*), etc.
Hybridisation must, however, be carried out every year using the original parents in order to avoid losing the advantages of heterosis, whence the price of F_1 seeds!

5 Mutations and Somations

We must, in the first instance, differentiate between these two terms of changes in characteristics.

a) When a change affects the embryo, in other words, when it appears after fertilisation, it is found among the descendants because it has been fixed on the chromosomes; this is a *germinal mutation.*

b) When the change involves only a vegetative portion of a plant it does not affect the embryo: this is a *gemmative mutation,* also known as a sport, a somatic mutation or a *somation.*

c) In all these cases, mutations and somations are noted only in isolated subjects and arise for no apparent reason.

5.1 Germinal mutation

There is a sudden change in the forms of a plant which will be repeated in its descendants.

Nevertheless, there should not be any confusion in respect of mutation, that is, the appearance for the first time of this new hereditary characteristic and the initial appearance of a non-apparent genotype, concealed by an extraordinary phenotype.

5.1.1 This may be due to a change in a gene.

If it is dominant, it is visible.

If it is recessive it is not apparent: many useful mutations must, in this way, have never been exploited.

5.1.2 It may be due to a change in the number of chromosomes in the nucleus: a chromosomal mutation, which is polyploidy.

Plants are, most often, diploid. This means that they have nuclei with $2n$ chromosomes.

When there is a chromosomal mutation, nuclei become triploid, with $3n$ chromosomes, or tetraploid, with $4n$ chromosomes, etc.

When a nucleus is triploid, the chromosomes are in odd numbers, resulting in a very considerable sterility among individuals. In such cases only one method of multiplication is possible: vegetative propagation.

When a nucleus is tetraploid it is very large; the cell is, therefore, larger and the plant is afflicted with gigantism.[8]

8 It is useful for the horticulturist to have tetraploid varieties: their fruits, flowers and leaves are larger, but it is difficult to cross triploid varieties (which have an odd number of chromosomes).
This had led to research by selectors on procedures that can result in germinal mutations:
- initiating changes using x-rays and, presently, with radioactive techniques,
- appearance of polyploidy due to cold treatment and, especially, with colchicine.

This is observed in the case of several plants that are provided to horticulturists: snapdragon (*Antirrhinum majus*), gaillardia (*Gaillardia grandiflora*), nemesia (*Nemesia strumosa*), china aster (*Callistephus chinensis*), phlox (*Phlox drummondii*), salsify (*Tragopogon porrifolius*), etc.

5.2 Somations or sports

5.2.1 Definition

In these, the reproductive organs are not involved.
There is only a sudden change in the buds and we understand that these new characteristics are not fixed on the chromosomes of the nuclei.
Somations are, therefore, not hereditary.[9]

9 Some somations can be useful in horticulture and the only way to fix the new characteristics is to adopt vegetative propagation. This is observed in the chrysanthemum (numerous sports). Also in the case of nectaries on the peach tree with downcovered fruits, the apple, the dahlia, the lily . . .
As in all descendants of individuals selected in this manner and propagated vegetatively, we have a *clone*.
A clone is produced from a rootstock.

5.2.2 Chimeras

The propagated stem has two types of tissue; it has characteristics of the parents as well as those that have undergone the somation: it is a chimera.
We denote as:
Periclinal chimera, one that has these characteristics all round the stem: dahlia, coleus.
Sectorial chimera, one that affects only a portion of the plant: sansevieria.
If the chimera is due to a somation following grafting we may, in very rare cases, witness the appearance of a new plant or portion of a plant known as a *graft hybrid*. The portion in question has two original tissues: one of the stock and the other of the scion.
Like a somation, this graft hybrid can under no circumstances be propagated by the sexual method.

6 Selection

We will not discuss natural selection, which applies to plants as it does to all living creatures, . . . where, in nature, only the fittest survives.
Man has, since long, adopted means of artificial selection.
This may be from an already existing population, without creating anything new—this is *conservative selection*—or from new populations—this is *creative selection*, from mutations, somations or hybridisations.
A study of Mendel's laws has taught us that, following a hybridisation,

— there are many new combinations,

— the "novelties" are not always visible immediately.

The selector must, therefore, be patient, be well organised, avoid dispersion of effort and have a keen sense of observation.
There are several techniques available to him.

6.1 Mechanical selection

Applying the principles of natural selection, the selector preserves only entire seeds that are the largest, heaviest, etc.[10]

10
Sorting and calibrating seeds is an indispensable mechanical selection method that will help to increase yields.

6.2 Mass or phenotypic selection

As the name implies, phenotypic selection takes into account only the "phenotype" of the selected variety.
In a given population, therefore, we need to:

— eliminate unsatisfactory individuals: those showing poor growth, poor flowering, poor fruiting, etc., and
— preserve only the "mass" that is most suited to our objectives to ensure the next generation.

This may be done in a single year or, better still, over several successive years.
It may be noted that fertilisation is a delicate matter in order to ensure continuity in the experiment. While it creates no problems for autogamous plants, which are self-fertilising, the same is not the case for allogamous plants, which must be protected in cross-fertilisation: for example, in isolated plantations.
Mass selection is also applicable in the case of plants that are propagated vegetatively.[11]

11
Mass or phenotypic selection:
— anemophilous plants: beetroot (*Beta vulgaris*);
— entomophilous plants: cabbage (*Brassica oleracea*), turnip (*Brassica napus*);
— plants propagated vegetatively: potato (*Solanum tuberosum*), rose (*Rosa canina*), perennial plants . . .

6.3 Genealogical selection or pedigree

As opposed to mass selection, which eliminates unsatisfactory individuals, here we have to

— extract a few individuals from a population that are considered to be the best of a selected lot,
— propagate them separately using this "head of the family",
— compare their descendants, eliminating undesirable lines.

Among other things, genealogical selection enables us to isolate a mutation or a somation in a homozygous population or to extract a "new variety" from a heterozygous population.

A comparison of lines obtained cannot be useful unless the multiplication has been conducted with a minimum of precautionary measures:

— care in fertilisation (especially cross-fertilisation),
— planting in soil of standard quality under identical technical conditions.[12]

6.4 Selection of healthy plants

This can be conservative or creative, using mass or genealogical selection techniques that assist us in providing horticulturists with healthy plants, free from diseases, or in the discovery of resistant cultivars.

12
This selection makes available to horticulturists:
— beans (*Phaseolus vulgaris*) without anthracnose,
— snapdragons (*Antirrhinum majus*) without rust,
— china asters (*Callistephus chinensis*) without fusarium,
— potatoes (*Solanum tuberosum*), strawberry plants (*Fragaria*) without virus diseases, etc.

Chapter 8

Systematics

1
Classification of Plants

1.1 Terminology

1.1.1 Systematics deals with the classification of plants

The first botanist to study this aspect was Linnaeus, who did so during the 18th century.
By classification we imply compartmentalisation.
For each compartment we need a label.
Precise terms are, therefore, used.
These terms have also been "codified".

1.1.2 Systematic classification

a) Plants (or individuals) with similar external or internal characteristics which can be propagated by crossings between them, resulting in fertile descendants, are covered in the same *species*.
The tomato (*Solanum lycopersicum*), which is used as a food, is one such example.
Let us be more specific.
b) Often, in the same species, differences (in shape, habit and colour of flowers . . .) are noted and individual plants can be propagated.
This is a *variety*, a term used for wild plants. If, however, the variety has been commercialised we prefer to use the term *cultivar*.
As examples, we have Marmande and Fournaise tomatoes.
We will, in the species, attempt to group plants on the basis of our common tomato (*Solanum lycopersicum*).
c) Several allied species that greatly resemble one another and have very many common characteristics form a *genus*.

For example, we have *Solanum* comprising *S. lycopersicum* (tomato), *S. tuberosum* (potato), *S. pseudo-capsicum* (love-apple plant), etc.

d) A *family* comprises a group of similar genera that have many common characteristics.

Thus, we have the family Solanaceae comprising genera such as *Solanum, Nicotiana* (tobacco) ...

Large families often need an intermediate classification: the sub-family.

It may be noted that the names of families generally end in "aceae"—Apiaceae, Oleaceae—and those of sub-families in "oideae"—Mimosoideae, Papilionoideae ...

e) A group of families that closely resemble one another form an *order* and the name usually ends in "ales".

For example, Solanaceae are included in the order of Tubuliflorales.

f) A combination of orders that have a certain number of common characteristics is known as a *class*.

This applies to the class of Dicotyledons, which includes: Tubuliflorales, Fagales, Rosales... *(see Table 8.2).*

g) Several classes that are clearly related form a *division.*

This applies to phanerogams comprising the dicotyledons and monocotyledons.

It may, nevertheless, be necessary to have subdivisions. In the division of phanerogams (spermatophytes) we find gymnosperms and angiosperms.

1.1.3 Nomenclature

A plant is always referred to in botany in *latin*, a universal language, and with a *binomial nomenclature*: the first name is written with a capital letter (name of the genus) and the second with a small letter (name of the species, known as the specific epithet), followed by an initial of the author of the name (for Linnaeus, it is L.).

Based on a decision taken by the International Botanical Congress held in Montreal in 1959 the specific epithet may, in exceptional cases, be written with a capital letter if it has been derived from a proper noun: name of a person, a place or a plant. In this book we have adopted this practice, which is currently accepted by French authors.

Depending on the particular case, a plant may be designated more specifically by the name of its variety, or even its cultivar, added to the genus/species binomial.

Names of varieties are written with a small letter or a capital letter depending upon the accepted rules for writing the names of species.
Names of cultivars are always given in inverted commas and are written with a capital letter.

1.2 General classification

What are the criteria for classification of plants in various compartments?
a) In the case of divisions: *on the presence or absence of flowers,*
b) In the case of sub-divisions:
— for non-flowering plants: *on the appearance of the vegetative apparatus,*
— for flowering plants: *on the appearance of the ovule.*
c) For classes of flowering plants: *on the structure of the embryo.*
d) For orders and families of flowering plants: *on the characteristics of the flower.*

1.3 Let us now examine, systematically, the characteristics of a flower

We have, while studying floral morphology, examined various organs and noted that we can identify:
— sepals,
— petals,
— stamens, and
— carpels.
These organs may be arranged in whorls (concentric circles) or in spirals.
We need to record what we have examined by a sketch.
We can have recourse either to a diagram or a section.

1.3.1 *A diagram* is a graphical representation of floral parts seen in transverse section.
From the outside inwards, we observe:
— the bract (or bracts), if any,
— the calyx,
— the corolla,
— the androecium,
— the pistil.
We note the precise arrangement of floral parts with respect to one another:

— whether they are joined together or separate,
— whether they cover each other,
— the arrangement of a whorl with respect to another.
We effectively complete the diagram with the floral formula, which is a record of what we have seen:
For example:
— number of sepals: 3S,
— number of petals: 3P,
— number of stamens: 6St,
— number of carpels: 3C
When the stamens or carpels are numerous and not in fixed numbers we record: indefinite = ∞
If the organs are joined we may:
— either add a small "j" to the indication;
— or write the term in parentheses: 3Cj or 3(C).

1.3.2 *A section* is a graphical representation of a flower based on a vertical section indicating the position in level of whorls with respect to one another. Conventionally, sepals are hatched and petals shaded.

1.3.3 Other graphical representations may be desirable:
— a section of a fruit,
— a section of an ovary, etc.

1.4 Classification of a herbarium

The principal rules that we have just noted and the systematic study of plants that we propose to undertake will help us to begin the classification of our herbarium. Family by family, and genus by genus.
For this purpose it is sufficient to proceed in an orderly manner and, initially, to use the following identification tables.

Table 8.1.

Divisions	Sub-divisions	Classes
Cryptogams Non-flowering plants	*Thallophytes*: No root, no stem, no leaf	Bacteria Fungi Algae Lichens
	Musci or Bryophytes: Stem and leaf only	Hepaticae Mosses
	Pteridophytes: or vascular cryptogams: Root, stem and leaf	Filicineae Equisetineae Lycopodineae

Divisions	Sub-divisions	Classes
Phanerogams Flowering plants	*Gymnosperms:* Naked ovules borne on certain scales as a result of which the seeds are not covered in the fruit	Natricae (no pollen tube)
		Vectricae (pollen tube)
	Angiosperms: Ovules covered in the ovary as a result of which the seeds are found inside the fruit	Monocotyledoneae seeds with a single cotyledon
		Dicotyledoneae seeds with two cotyledons

2 Thallophytes

This sub-division includes several classes that may be identified in the following manner:

Thallophytes	Without chlorophyll	Unicellular plants	Bacteria
		Plants with one or several cells	Fungi
	With chlorophyll —1 or several cells		Algae
	Symbiotic plant —1 alga + 1 fungus		Lichens

We will study only those thallophytes that are of agricultural interest.

2.1 *Bacteria*

These are microscopic organisms 1/100 to 1/1000 mm in size.

Bacteria have various forms: round, rodlike or spiral.

They are propagated:

— either by division: rapid multiplication in a favourable environment;
— or by spores in favourable conditions; this is dormancy.

Bacteria may release heat.[1]

1 Thermogenic bacteria enable the "rise of warm layers" from organic matter (*see the course material on General Horticulture*).

Bacteria may exist in the presence of oxygen (air) or without oxygen. They are accordingly known as aerobic or anaerobic bacteria.

Some of them play a primordial part in the nitrogen-cycle process in the soil.

Micrococcus urae transforms urea into ammonium carbonate.

Micrococcus nitrifians produces nitrous acid from ammonia.

2.2 Fungi

- Fungi
  - Protoplasm without a wall causing the fungus to look like a jelly → Myxomycetes
  - Protoplasm covered by a wall
    - Non-partitioned thallus forming a single and branched tube → Oomycetes or Siphomycetes
    - Partitioned thallus
      - Spores in asci → Ascomycetes
      - Spores in basidia, the reproductive apparatus in the shape of a cap → Basidiomycetes

This classification is primarily based on anatomical criteria and subsequently on the reproductive mode.[2]

2 We will leave it to the course material on parasitology to deal with this important aspect of horticulture and confine ourselves to a few typical examples for each group.

We find in:

1) Myxomycetes: Plasmodiophora brassicae (finger and toe in cabbage). Reproduction through sporangia containing spores.

2) Siphomycetes: Phytophthora infestans (late blight in the potato). Reproduction by spores contained in conidia (during growth) and also by zygotes (for preservation of the species in winter). This method involves sexual multiplication.

This order comprises mainly:
— the mucorinales, which include the moulds,
— and the peronosporales, which include mildews and white rusts.

3) The Ascomycetes comprise:

— the discomycetes: morels, yeast (production of beer, wine . . .),

— the perisporiaceae: *Penicillium* (penicillin), *Aspergillus* (for making Roquefort cheese), truffles, oidium on the peach tree;

— the pyrenomycetes: ergot of rye.

The use of yeasts is well known in industries producing beer, cider and wines.

Some perisporiaceae are essential for the manufacture of antibiotics.

4) Basidiomycetes.[3]

There are three principal families:

— the Agaricaceae: the edible mushroom, the cantharellus and the amanita;

— the Polyporaceae: polyporus (on the trunks of trees), boletus and the penny-bun mushroom;

— the Lycoperdaceae: the puff-ball fungus.

3
This last group of fungi comprises most of the edible species that exist in nature or are grown commercially.

2.3 Algae

These plants exist mainly in damp regions or in water (fresh or sea water) and are classified depending upon the colour of the supernumerary pigment that accompanies the chlorophyll.

This enables us to prepare the following table:

Algae	Containing only chlorophyll		Chlorophyceae: green algae
	Containing colouring pigments that alter the green colour of the chlorophyll	Yellow, brown pigment, reproductive cells with cilia	Pheophyceae: brown algae
		Red pigment, no cells with cilia	Rhodophyceae: red algae
		Blue pigment	Cyanophyceae: blue algae

2.4 Lichens

Some, such as the wall parmelia, have a leafy appearance.

Others have a tendency to ramify.

There are also others with a soft consistency.

All of them have considerable vitality and are highly resistant to harsh conditions of growth.

3 Mosses or Bryophytes

These sub-divisions comprise two classes:
— the mosses
— the hepatics.
The class of hepatics is intermediate between the sub-division of thallophytes and the class of actual mosses. In both classes the reproductive mode is very similar. These plants are not of much agricultural interest.

4 Pteridophytes

We know that these cryptogams have roots, stems and leaves.
The appearance of the stem enables a classification of Pteridophytes.

Unbranched stem		*Filicineae* ferns
Branched stem	Branches in whorls	*Equisetineae* horsetails
	Branches of a dichotomous type	*Lycopodineae* selaginella (*Selaginella cuspidata*)

4.1 Filicineae: *ferns*

This family with many species (several thousands) is characterised by:
— numerous fine roots;
— an underground stem (rhizome), therefore, they are perennial species;

— crosier-shaped leaves when they are young, developing later as "fronds".

These fronds are of varying sizes (even as long as several metres) and, often, highly denticulate.

In tropical regions, ferns sometimes have "trunks" and resemble palm trees; on these trunks we can observe the scars of fronds that have been shed.

Ferns are generally found in humid areas, sheltered from sunlight.

They may exist as "epiphytes".[4]

4 Many species are grown by horticulturists: *Adiantum, Aspidium, Nephrolepis, Platycerium, Polypodium, Pteris,* etc., all of which have different uses: cut leaves, potted plants, decorative vases, winter gardens . . .

4.2 Equisetineae: *horsetails*

Horsetails are, par excellence, typical plants of marshlands and humid soils.

Their horticultural utility is limited to recognition of an environment based on its flora.

4.3 Lycopodineae: *selaginella*

Only the selaginella is of horticultural interest to floriculturists, who grow one of its species that has a somewhat silvery appearance (*see a description of this in Morphology*).

5 Gymnosperms

Let us refer to the definition of this first sub-division of phanerogams:

"Flowering plants, with naked ovules, borne on open scales, as a result of which the seeds are not contained in a fruit."

The existence or absence of a pollen tube during fertilisation enables an initial classification.

If there is no pollen tube, these are natrices: Cycadaceae, Ginkgoaceae. If there are pollen tubes, these are vectrices: Conifers.

The most important order from the point of view of horticulture is that of the conifers.

5.1 Conifers[5]

The conical stem thickens every year. The wood contains secretory resin canals that develop even within the leaves.

The leaves persist for several years and are rather thin, often in the form of needles (pine, *Pinus sylvestris*) and even in the form of scales (Thuja of Canada, *Thuja occidentalis*). In the spring new leaves appear: the tree always remains green.

5.2 Cycadaceae *and* Ginkgoaceae

These two groups, which are very similar to the conifers, are of limited interest in horticulture.

Cycadaceae, with unbranched stems, resemble palm trees.

5

This order comprises many species cultivated in ornamental nurseries:

- Abietaceae, with conical inflorescences.
  Abies: Fir tree: with "self-shedding" cones as a result of which there are no dry cones on the tree.
  Cedrus: Cedar tree: needles arranged in bunches.
  Picea: Spruce-fir: persistent cone in a single piece that remains hanging on the branch for a fairly long time.
  Pinus: Pine: needles inserted in groups of 2, 3, 4 and 5.
  Araucaria: Araucaria: imbricate leaves.
  Larix: Larch: conifer with a deciduous leaf.
- Cupressaceae: cupressus with a spherical cone or strobilus.
  Cupressus, *Chamaecyparis* (Cypress): with a small cone.
  Juniperus (Juniper): cone with fleshy scales in the form of a berry. Thuja.
- Taxaceae (Taxus)
  Yew tree: conifer without a cone, with isolated female flowers producing fruits in the shape of a cap or "arils". Many variegated varieties exist among several of these species.

6 Angiosperms

Angiosperms are flowering plants with ovules enclosed in an ovary and, therefore, the seeds are contained in a fruit.

This sub-division is, definitely, most important in horticulture both from the quantitative (number of species) and the qualitative (commercial importance) points of view.

This, perhaps, explains the complications in classification as under:

Class	Di- or Monocotyledons
Sub-class	Gamo-, A- or Dialypetalae
Orders	Numerous
Families	Even more numerous

Table 8.2 shows some of these families classified with respect to each other. It is not necessary to stress that this table provides only a restricted idea of all the plant families in angiosperms.

It may be noted that this has been based partly on Engler's classification as adopted by Guillaumin in the *Bon Jardinier* (Good Gardener), 152nd edition.

Our aim is to study some of these families as examples in order to familiarise ourselves with a systematical analysis of one of them and we will deliberately restrict ourselves to 14 examples.

Each study will comprise:

— the vegetative characteristics of the family,
— an analysis of the flower,
— the floral diagram and formula,
— the horticultural utility of the family.

} All based on a typical example of the family

6.1 Dicotyledons

This class comprises plants with:

■ *morphologically*:
— two cotyledons in their seeds,
— floral parts, four or five or in multiples of four or five,
— a non-petaloid calyx,
— leaves with branched veins.

■ *anatomically*: with secondary growths.

Nevertheless, it should be borne in mind that in systematics, as in other matters, exceptions prove the rule.

6.1.1 Dialypetalous dicotyledons

Flowers of the group have petals that are free from each other.

6.1.1.1 Ranunculaceae

Type: ranunculus or buttercup *(Ranunculus asiaticus)*

Table 8.2.

Class	Sub-class				Order	Family
Dicotyledons	Dialypetalae	Ovary free	Stamens free	Axillary placentation	Ranales	*Ranunculaceae*, Nymphaeaceae
					Geraniales	*Euphorbiaceae*
					Sapindales	Buxaceae, Hippocastanaceae, Celastraceae
				Parietal placentation	Rhoeales	*Cruciferae*, Papaveraceae
			Stamens joined at least at the base		Rosales	*Leguminosae*, *Rosaceae*, Crassulaceae
		Ovary adherent			Umbellales	*Umbelliferae*, Araliaceae
					Opuntiales	Cactaceae
	Apetalae	Ovary free	Unisexual flowers		Urticales	Urticaceae
			Bisexual flowers		Polygonales	Polygonaceae
		Ovary adherent			Fagales	Cupuliferae
	Gamopetalae	Ovary free	Pistil with more than 2 carpels		Primulales	Primulaceae
					Tubuliflorales	*Solanaceae*, Convolvulaceae, *Labiatae*, Boraginaceae, Acanthaceae, Scrophulariaceae
			Pistil with two carpels		Gentianales	Oleaceae

			Order	Family
	Ovary adherent		Campanulales	*Compositae* Campanulaceae
			Cucurbitales	Cucurbitaceae
			Rubiales	Rubiaceae Caprifoliaceae
Monocotyledons	Regular flowers with a perianth	Seed with fleshy endosperm	Liliales	*Liliaceae* Iridaceae Amaryllidaceae
		Seed with farinaceous endosperm	Farinales	Bromeliaceae
	Irregular flowers		Microspermae	*Orchidaceae*
	Young flowers surrounded with bracts		Palmales	*Palmae*
	Flowers without a aperianth		Graminales	*Gramineae* Cyperaceae

Note: Only the families whose names are printed in italics will be dealt with in this book.

General characteristics

Most of the Ranunculaceae are herbs; sometimes climbing shrubs such as the clematis.

The leaves are alternate, sheathing the stem.

The flowers are symmetrical with respect to the axis of the pedicel (they are said to be regular) and have:

— five green sepals, free from each other,

— five golden yellow petals,

— a variable number of stamens; they are external or lateral and free,

— the pistil is formed from a large number of carpels with an axillary placentation,

— the floral formula is: 5S, 5P, ∞ St, ∞ C.

The fruit is a collection of achenes.

The ranunculus and clematis with several carpels bear a fruit called a polyachene (each carpel is an achene).

The aquilegia or columbine has five carpels, each of which contains many ovules: each carpel produces a follicle.[6]

6

- Numerous carpels with a single ovule
  - a) No corolla
    clematis (Clematis montana), anemone (Anemone coronaria).
  - b) With corolla
    ranunculus (Ranunculus asiaticus),
    spring adonis (Adonis vernalis).
- Few carpels with several or many ovules
  - a) Extrorse anthers (slit turned outwards).
    aconite (Aconitum napellus),
    aquilegia (Aquilegia caerulea),
    rocket larkspur (Delphinium ajacis),
    hellebore (Helleborus abschasecus).
  - b) With lateral anthers.
    peony (Paeonia lactiflora).

Many of these are ornamental plants including species with numerous stamens producing beautiful double flowers (peony, etc.) as a result of a transformation of stamens into petals.

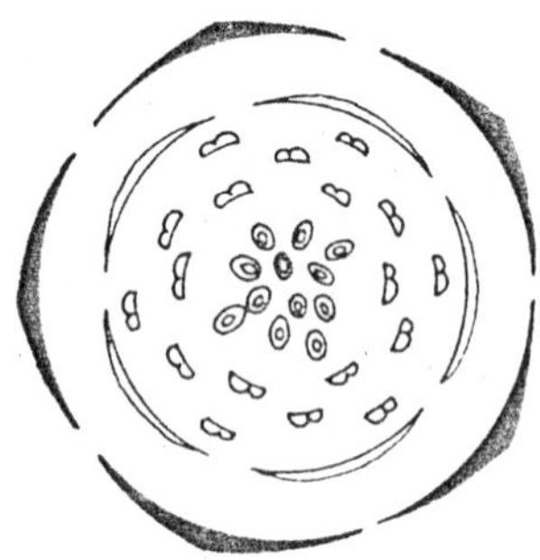

Fig. 8.1. Floral diagram of ranunculus

6.1.1.2 Euphorbiaceae

This is a highly heterogeneous family; the flower may have both a calyx and a corolla, or only a calyx and may even be without a perianth.

Type: castor (*Ricinus communis*)

General characteristics

Leaves alternate, simple.

The vegetative parts have latex vessels that contain a white liquid which turns black on contact with air.

The unisexual flowers are regular, grouped in clusters of cymes (the lower cymes are formed of male flowers and the upper cymes of female flowers).

— The male flowers have five sepals that are separate from each other and no petals. The stamens are branched.

— The female flowers have three separate sepals; the ovary has three carpels.

Floral formula: 5S 0P, ∞ St and 3S, 0P, 3Cj.

The fruits are capsules containing large seeds with an oleaginous endosperm.[7]

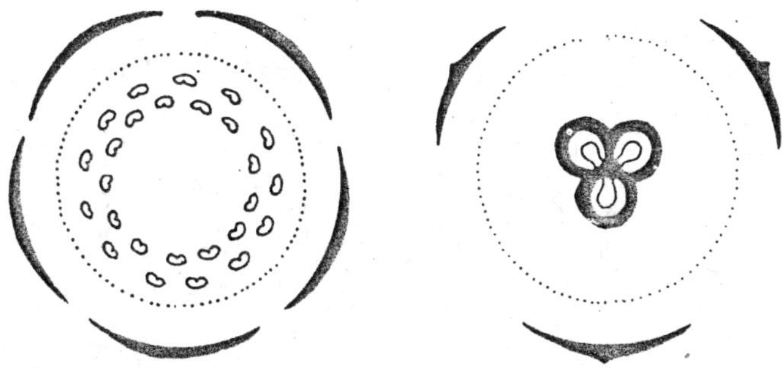

male flower female flower

Fig. 8.2. Floral diagrams of the castor plant

6.1.1.3 Cruciferae

A very homogeneous family in which examination of a single plant enables a study of the organisation of the entire family.

Plants of this family are herbaceous.

Type: wallflower (*Cheiranthus cheiri*).

General characteristics

The wallflower is an annual plant.

The roots have two vascular strands and four rows of lateral roots.

The roots are tap roots.

The leaves are alternate without a sheath or stipule.

The flowers, in simple racemes, have a somewhat regular covering:

— four violet-tinted sepals that are free from each other and slightly unequal in size;

— four petals, unequal in size, and free from each other in the shape of a cross (whence the name of the family);

— six stamens of unequal size in two whorls, four are long and form one whorl and the other two, located a little lower down and shorter, form the second whorl;

— the pistil, with two carpels, has a parietal placentation.

— The floral formula is 4S, 4P, 6St (2 small), 2C.

7

This family comprises monoecious and dioeceous plants:

— Euphorbia (Euphorbia pulcherrima),

— Hevea (Hevea brasiliensis),

— Croton (Codiaeum variegatum).

The fruit is a siliqua; if it were less than three times longer than broad it would be called a silicula (shepherd's purse).
The fruit of the *Raphanus sativus* (radish) is composed of successive swellings.[8]

8
A few such food plants are: cabbage (*Brassica oleracea*), mustard (*Sinapis nigra*), colza, water cress (*Nasturtium officinale*), radish (*Raphanus sativus*).
Some ornamental plants: wallflower (*Cheiranthus cheiri*), penny flower (*Lunaria annua*), rock alyssum (*Alyssum saxatile*), sweet alyssum (*Iberis sempervirens*).

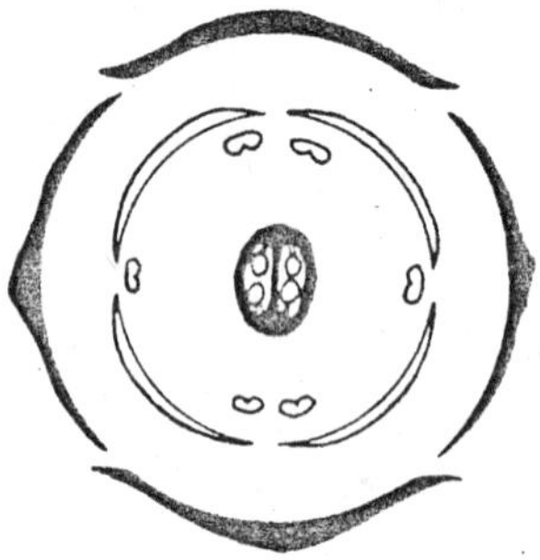

Fig. 8.3. Floral diagram of the wallflower

6.1.1.4 Leguminosae

These represent one of the most important and fairly heterogeneous families.
Type: the pea (*Pisum sativum*).
a) General characteristics
Herbaceous plants.
The leaves are compound and alternate; some are transformed into tendrils.

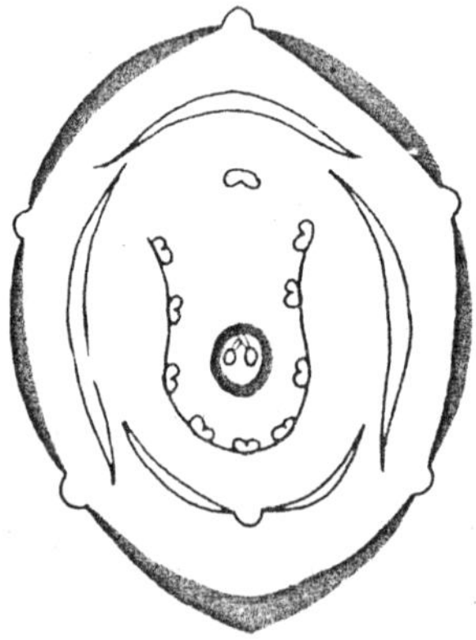

Fig. 8.4. Floral diagram of a pea plant

The flowers have a bilateral symmetry and comprise:
— five joined sepals,
— five petals of unequal size, fixed to the base of the sepals and differentiated in a standard, two wings

and two carinas,
— ten stamens, nine of which are fused at the base by thin filaments, and one is free,
— a free pistil with one carpel.
— The floral formula is (5S), 5P, 10St (of which 9 are j), 1C.

The fruit is a pod or a legume, whence the name of the family.

b) Botanical classification

There are three large sub-families because of the diversity in types of the same family:

■ *Papilionaceous-type corolla:* example: the pea (*Pisum sativum*). Sub-family Papilionoideae.

■ *Irregular corolla,* but not of the papilionaceous type. Example: Judas tree (*Cercis siliquastrum*). Sub-family Cesalpinioideae.

Flowers with joined sepals, the standard is covered by the wings and the carina that is free covers the wings. The fruit is a pod.

■ *Regular corolla:* sensitive plant (*Mimosa pudica*), sub-family of the Mimosoideae.

The flowers are regular with four sepals, four petals and four stamens.[9]

9

Sub-families Mimosoideae with the mimosa (*Acacia dealbata*), the *Cesalpinioideae* with the Judas tree (*Cercis siliquastrum*) and the *Gleditschia triacanthos* are ornamental plants.

The sub-family Papillionoideae includes:

■ Ornamental plants:
Genista (*Genista scoparia*)
Cytisus (*Cytisus laburnum*)
Lupine (*Lupinus polyphyllus*)
Robinia (*Robinia pseudo-acacia*)
Sophora (*Sophora japonica*)
Wistaria (*Wistaria sinensis*), etc.

■ Fodder plants:
Bird's-foot trefoil (*Lotus*)
Clover (*Trifolium*)
Lucerne (*Medicago*)
Vicia (*Vicia*)
Sainfoin (*Onobrychis*), etc.

■ Plants used as food by man:
Bean (*Phaseolus vulgaris*)
Soya (*Soja hispida*)
Pea (*Pisum sativum*)
Lentil (*Ervum lens*)
Broad bean (*Faba vulgaris*), etc.

6.1.1.5 Rosaceae

This is a highly heterogeneous family due to differences in the floral structure and morphological aspects.

Type: the strawberry (*Fragaria vesca*).

General characteristics

A herbaceous plant.

Flowers regular with five green sepals and five white petals inserted at the base of the sepals.
— Numerous stamens also inserted at the base of the sepals and free from one another,
— Numerous independent carpels,
— Floral formula: 5S, 5P, ∞ St, ∞ C, with five small additional petals forming a calycule or epicalyx.

The fruits are achenes arranged on the fleshy receptacle.[10] (see p.194)

6.1.1.6 Umbelliferae

A very homogeneous family.

Type: the carrot (*Daucus carota*).

General characteristics

A herbaceous plant.

The stems are hollow and erect.

The leaves are alternate with wide sheaths at the base surrounding the stem; they are highly denticulate.
Carrot flowers are small and white; they are grouped in compound umbels; each umbel has a series of bracts at the base forming an involucel and the composite umbel has an involucre.
Each flower comprises:
— five very small sepals, joined to an ovary of irregular dimensions,
— five petals, folded back at the tips,
— five stamens,
— the ovary formed from two carpels and surmounted by two styles.
— The floral formula is 5S, 5P, 5St, 2C.

The fruit is a diachene or cremocarp with secretory canals and the two mericarps separating at maturity.[11]

10 (p.193)

These depend on the appearance of the fruits and characteristic of the carpels. The main tribes can be classified as follows:

			Tribe / plant	Species
Carpels do not adhere to the calyx	Fruit is a drupe		Amydaleae	(only one carpel)
			Almond	*Amydalus vulgaris*
			Plum	*Prunus domestica*
			Peach	*Amygdalus persica*
			Apricot	*Prunus armeniaca*
			Cherry	*Prunus avium*, etc.
	Fruit is a follicle		Spireae	(only 5 carpels)
			Spiraea	Spiraea
			Kerria	*Kerria japonica*, etc.
	The fruits are achenes	On a convex receptacle	Fragareae	
			Strawberry	*Fragiara vesca*
			Potentille	*Potentille*
			Raspberry	*Rubus idaeus*, etc.
		On a concave receptacle	Rosaceae	
			Rose	*Rosa*, etc.
Carpels adhere to the calyx			Pomaceae	
			Pear	*Pyrus communis*
			Apple	*Malus communis*
			Sorbus	*Sorbus aucuparia*
			Hawthorn	*Crataegus oxyacantha*
			Cotoneaster	*Cotoneaster franchetii*
			Quince	*Cydonia japonica*

The ornamental and especially the fruit value of this family is important in horticulture.

11

- Tuberous roots used as foods: carrot (*Daucus carota*), parsnip (*Pastinaca sativa*).
- Leafy vegetables: celery (*Apium graveolens*),
- The other plants are aromatic: anise (*Pimpinella anisum*), parsley (*Petroselinum sativum*), chervil (*Cerefolium sativum*), angelica (*Archangelica officinalis*), fennel (*Foeniculum dulce*).

6.1.2 Apetalous dicotyledons

Flowers of this group have no petals.

6.1.2.1 Cupuliferae

Type: English oak (*Quercus robur*).

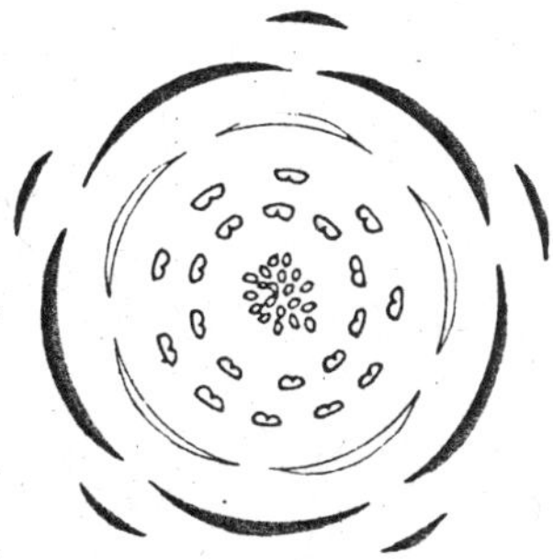

Fig. 8.5. Floral diagram of the strawberry

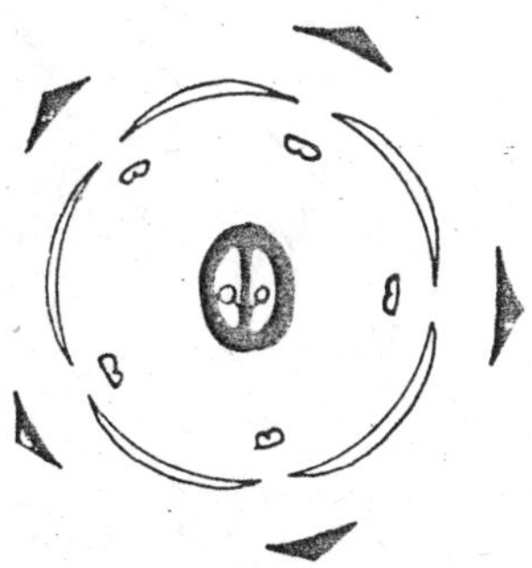

Fig. 8.6. Floral diagram of the carrot

General characteristics

The leaves are alternate, deciduous and simple.

The unisexual flowers are borne on the same tree and are grouped in "catkins". Male catkins are longer and have more flowers than female catkins.

— Male flowers have, on the average, 6 sepals and 5 to 10 stamens.

— Female flowers are formed of ten sepals joined to the ovary and 3 united carpels with 2 ovules in each loculus.

These female flowers are inserted in a small cup: the cupule, whence the name of the family.

— The floral formulae are 6S, 5 to 10 St, and 6S, 3C.

The fruits are nuts: of the three possible ones in each cupule only a single one remains and the entire structure forms the acorn.[12]

12

The principal groups are classified by the presence or absence of a calyx in the flowers.

- Fagaceae: male and female flowers with a calyx:
  Oak: *Quercus robur, Q. sessiliflora,*
  Beech: *Fagus sylvatica,*
  Chestnut: *Castanea vulgaris.*
- Betulaceae: calyx in male flowers or in female flowers:
  Coryleae: calyx in male flowers:
  Hazel: *Corylus avellana,*
  Hornbeam: *Carpinus betulus.*
  Betuleae: calyx in female flowers:
  Birch: *Betula verrucosa,*
  Alder: *Alnus cordifolia.*

The family is a limited one but very important economically because it includes the main indigenous forest species.

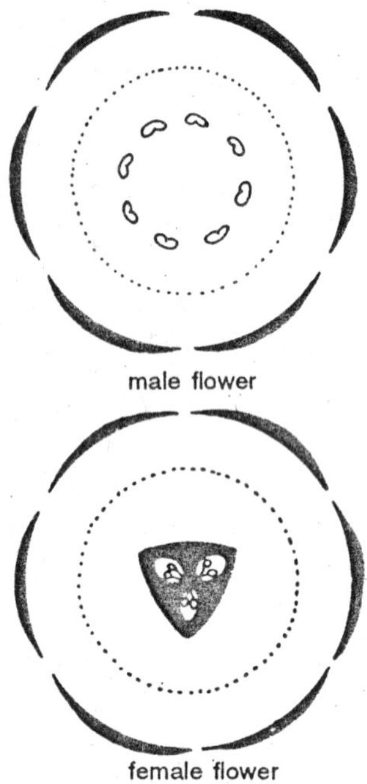

Fig. 8.7. Floral diagrams of the oak

6.1.3 Gamopetalous dicotyledons

The flowers in this group have petals that are, to some extent, united.

6.1.3.1 Solanaceae

The family is fairly homogeneous.
Type: the potato (*Solanum tuberosum*).

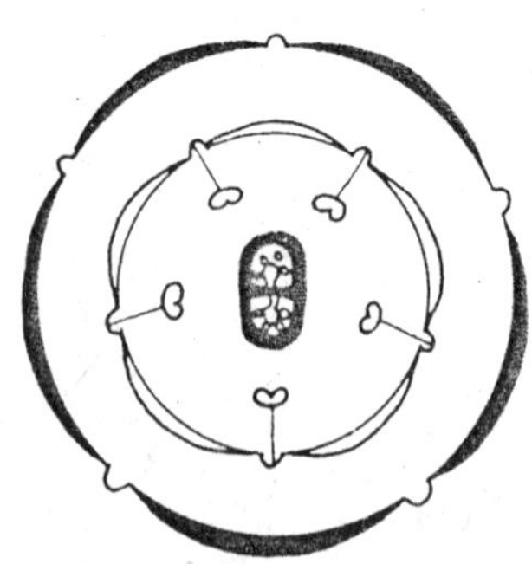

Fig. 8.8. Floral diagram of the potato

General characteristics

The leaves are compound and alternate.

An aerial stem and an underground stem with swellings (tubers).

Regular flowers with: five green united sepals, five whitish pink petals, five stamens attached to the petals, a pistil with two carpels, and a single stigma.

Floral formula: 5Sj, 5Pj, 5Stj, 2C.[13]

6.1.3.2 Labiatae

This family is very homogeneous.

Type: the white dead nettle (*Lamium album*), improperly called the white nettle.

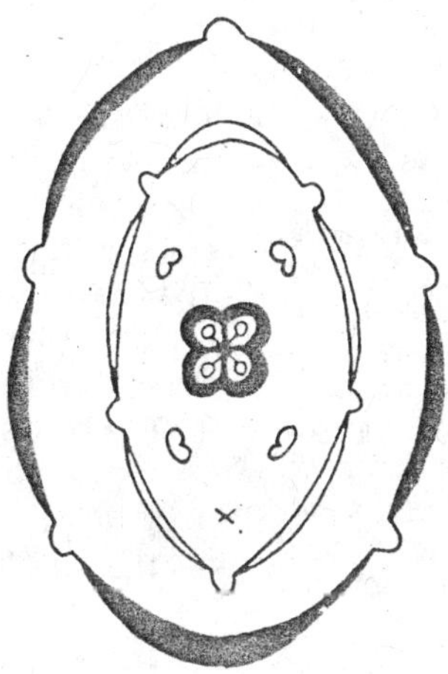

Fig. 8.9. Floral diagram of the white dead nettle

General characteristics

Leaves simple, opposite, pubescent.

The stem is square in section (angles reinforced by collenchyma).

The flower is irregular, with a bilateral symmetry, and tubular, bilabiate (whence the name of the family): there are five green united sepals, five petals united in the form of a tube (one lip is due to the union of two petals and the other of three petals), four stamens of which two are longer and an ovary with two carpels each divided into two distinct parts.

The fruit is a tetrachene or carcerulus with four nutlets.

Floral formula: (5S), (5P), 4St, (2C).[14]

13

- Many plants used as foods:
  - Potato: *Solanum tuberosum*
  - Tomato: *Solanum lycopersicum*
  - Aubergine: *Solanum melongena*
  - Capsicum: *Capsicum annuum*
- Industrial plants
  - Tobacco: *Nicotiana tabacum*
- Ornamental plants:
  - Flowering tobacco: *Nicotiana affinis*
  - Petunia: *Petunia grandiflora*
  - Schizanthus: *Schizanthus wisetonensis*
  - Salpiglossus: *Salpiglossus sinuata*
  - Datura: *Datura stramonium*

Various Solanaceae members contain alkaloids:

- Tobacco: *Nicotiana affinis:* nicotine
- Datura: *Datura stramonium:* daturine
- Belladona: *Atropa belladonna:* atropine
- Strychnos: *Strychnos nux-vomica:* strychnine, curare, etc.

14

Nearly all plants of this family contain many essential oils that are extracted by distillation.

These plants are naturally very fragrant: they are aromatic plants.

Lavender: *Lavandula spica*

Mint: *Mentha viridis*

Sage: *Salvia officinalis*

Wild thyme: *Thymus serpyllum*

Rosemary: *Rosmarinus officinalis*

Thyme: *Thymus vulgaris*

A few species are used as foods:

Japanese artichoke: *Stachys tuberifera*

Some are ornamental:

Sage: *Salvia officinalis*

Fig. 8.10. Floral diagram of Compositae

6.1.3.3 Compositae

This is the largest family.

General characteristics

Each capitulum is an inflorescence composed of several flowers or florets.

The flowers are placed close together in a type of receptacle formed by a terminal widening of the stem like a head.

— The bracts, with respect to the capitula, play the same role as do the sepals with regard to the flower.

— The sepals are reduced to a few hair-like structures.

— There are five petals joined in a tube that form flowers of a tubular shape (thistle) or flowers of a ligulate shape: ligules (dandelion).

— Five stamens with free filaments, but the anthers are united, and filaments are attached to the petals. Anthers form a tube.

— Young flowers have anthers that are longer than the stigmas.

In adult flowers the stigmas rise, pass within the tube formed by the united anthers and collect pollen.

— There are two carpels, united with each other, but only a single ovule develops.

— Floral formula: 0S, 5Pj, 5Stj, 2Cj.

The fruit is an achene surmounted with a tuft of hairs (persistent calyx).[15]

In the tissues there are secretory and laticiferous canals (essential oils and latex).

15

Classification

- **All flowers with tubular corolla, Tubiflorae**
  Thistle: ***Carduus nutans***
  Cornflower: ***Centaurea cyanus***
  Artichoke: ***Cynara scolymus***
- **Flowers towards the centre with tubular corolla and towards the periphery with ligulate corolla, rayed:**
  Ox-eye daisy: ***Leucanthemum vulgare***
  Chrysanthemum: ***Chrysanthemum indicum***
  Cineraria: ***Seneci cineraria***
  Marigold: ***Tagetes patula***
  Dahlia: ***Dahlia hortensis***
  Aster: ***Aster amellus***
  Daisy: ***Bellis perennis***
- **All flowers ligulate**
  Chicory: ***Cichorium intybus***
  Lettuce: ***Lactuca sativa***
  Dandelion: ***Taraxacum dens leonis***
  Salsify: ***Tragopogon porrifolius***

Ornamental plants of Compositae can easily produce "double" flowers by transformation of tubular flowers in the capitulum into ligulate flowers: chrysanthemums, dahlias, etc.

6.2 Monocotyledons

This class includes plants that have:

■ *Morphologically:*

One cotyledon in the seed, floral parts in threes or multiples of three, a petaloid calyx, leaves without a petiole and with parallel veins.

■ *Anatomically:*

Several concentric circles of vascular bundles; and secondary growth in the roots and stems.

6.2.1 Liliaceae

A large family.

Heterogeneous in respect of the vegetative parts and fairly homogeneous with regard to the flower.

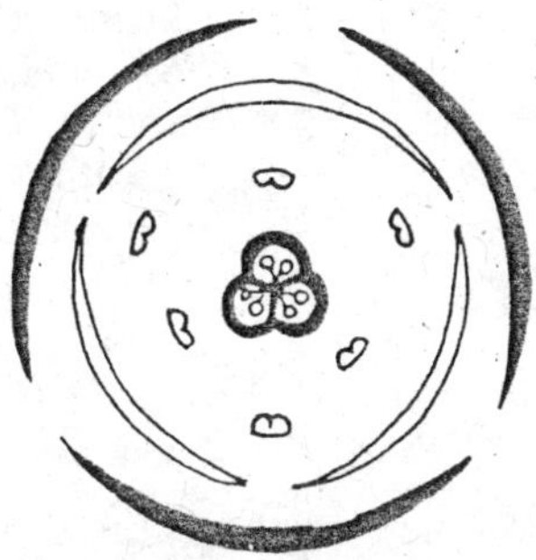

Fig. 8.11. Floral diagram of the lily

Type: Madonna lily (*Lilium candidum*).

General characteristics

A perennial plant due to its scaly bulb.

The leaves are elongate with parallel veins.

Flowers in racemes comprise: three free sepals and three free petals.

— There are three united carpels and a superior ovary.

— Six introrse stamens in two whorls.

— The sepals and petals are of the same colour.

Floral formula: 3S, 3P, 6St, 3Cj.

The fruit is a capsule.[16]

6.2.2 Orchidaceae

The family comprises the largest number of plants belonging to the class of monocotyledons.

Their characteristics are extremely uniform.

16

Principal groups

■ Bulbous plants: tulips

Ornamental plants:

Lily: *Lilium candidum*

Fritillaria: *Fritillaria imperialis*

Tulip: *Tulipa*

Scilla: *Scilla peruviana*

Hyacinth: *Hyacinthus orientalis*

Muscari: *Muscari moschatum*

Plants used as food:

Garlic: *Allium sativum*

Shallot: *Allium ascalonicum*

Onion: *Allium cepa*

Leek: *Allium porrum*

■ Rhizomatous plants (fruits in a berry): Asparagineae.

Asparagus: *Asparagus officinalis*

Lily of the valley: *Convallaria majalis*

Butcher's broom: *Ruscus aculeatus*

Aspidistra: *Aspidistra elatior*

■ Trees with erect stems: Dracaeneae

Dragon dracaena: *Dracaena draco*

Yucca: *Yucca filamentosa*

Aloe: *Aloe vera*

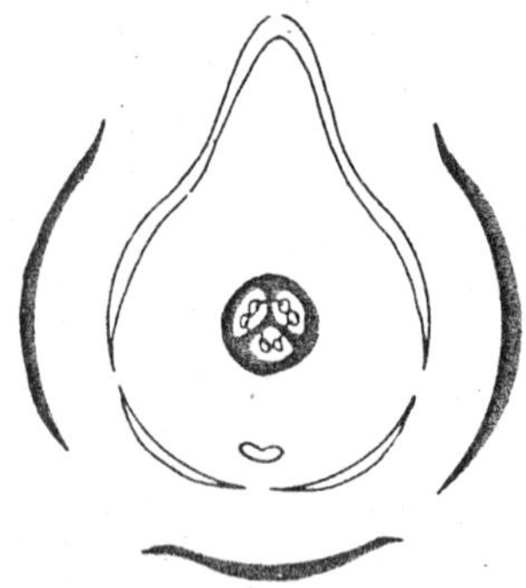

Fig. 8.12. Floral diagram of an orchids

a) General characteristics

European orchid species are mainly terrestrial, while those of equatorial regions are:

epiphytes (living attached to trees that provide them support),

saprophytes (living on decomposing matter).

The root system comprises:

— thin roots,

— two tuberous roots, one wrinkled, the other smooth.

Some orchids are climbers.

b) Reproductive parts

Type: spotted orchis (*Orchis maculata*).

The leaves are alternate at the base of the floral spike.

The symmetry of the flower is bilateral.

— The three sepals are petaloid (they resemble petals closely) and are tinted pink and white.

— The petals comprise two petals arranged symmetrically on either side of the upper sepals forming a type of cap and a third petal, larger than all other parts of the perianth, in the form of a spur at the lower side of the flower and called a labellum.

— The androecium is composed of a single stamen with a sessile anther.[17]

The pollen grains are combined in a club-like structure called a pollinium. There are two pollinia per stamen. The base of a pollinium is sticky and this helps in entomophilous pollination as the pollinia are attached to the heads of insects when they "visit" the flowers.

— The pistil comprises three united carpels with a single loculus.

The fruit is a capsule.

Floral formula: 3S, 3P, 1St, 3Cj.

17

As an exception to the general rule for the family, in the *Cypripedium* there are two or three stamens. The other Orchidaceae with a single stamen are plants with magnificent ornamental flowers with a rich structure and colour:

orchis, vanda, cattleya, odontoglossum, dendrobium, etc.

Specialist horticulturists have obtained many hybrids.

c) Physiology of Orchidaceae

Noël Bernard made a detailed study of the special development in these plants.

Let us follow up this development, which can only be achieved by the use of a fungus of the *Rhizoctonia* type that lives in symbiosis with these plants.

■ *Stage 1.* The symbiotic fungus enters the seed and it is only from this moment that the seed can germinate. The embryo produces a type of tuber infested with the fungus that develops a bud, which is the origin of the first tuberous root.

■ *Stage 2.* At the end of the initial year of growth in July, the first tuberous root separates from the infested tuber: the latter, having exhausted its reserves, wilts and the plant is, for a while, not infested and produces normal vegetative organs: roots and leaves.

■ *Stage 3.* In October of the same year the plant is, again, infested and does not grow normal roots or leaves but a second tuberous root.

Conclusion. The development of Orchidaceae is alternate and there are two distinct periods:

— A period of isolation during which growth is normal with single roots, leaves and flowers while the plant is free of any symbiosis with a fungus.

— A period of infection during which a tuberous root develops that accumulates reserve material which can be used later on.

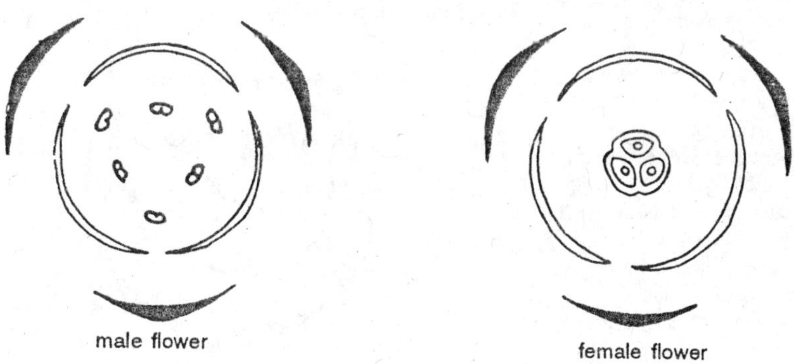

Fig. 8.13. Floral diagram of the Phoenix

6.2.3 Palms

Plants of this family are found mainly in tropical regions.

General characteristics

The stem is a caudex ending in a cluster of leaves that are highly denticulate and may be pinnate: date

(*Phoenix dactylifera*) or palmate (*Chamaerops excelsa*).
The unisexual flowers are borne on mono- or dioecious trees, combined in clusters that emerge from the base of the leaves.
Floral formulae: 3S, 3P, 6St, and 3S, Sp, 3Cj (a single ovule of three remains).
The fruit is a berry or a drupe.[18]

6.2.4 Gramineae

The family is highly homogeneous.
Type: wheat (*Triticum sativum*).
General characteristics
Fasciculate roots, not very deep.
Stem (or stalk) unbranched, hollow, erect, rigid, cylindrical, swollen at the nodes and solid at these nodes.
The leaves are alternate, inserted at the surface of the stem in two opposite rows.
The flowers are grouped in a compound spike. Secondary spikes are called spikelets. At the base of each spikelet there are two glumes (protection), two glumelles or flowering glumes (the bracts) and two glumellules or lodicules (the sepals).
There are no petals.
The flower has the following essential parts:
— three pendulous stamens (3St),
— a rounded ovary with one carpel (1C).
The fruit (grain of wheat) is a caryopsis.
The endosperm is farinaceous (starch).[19]

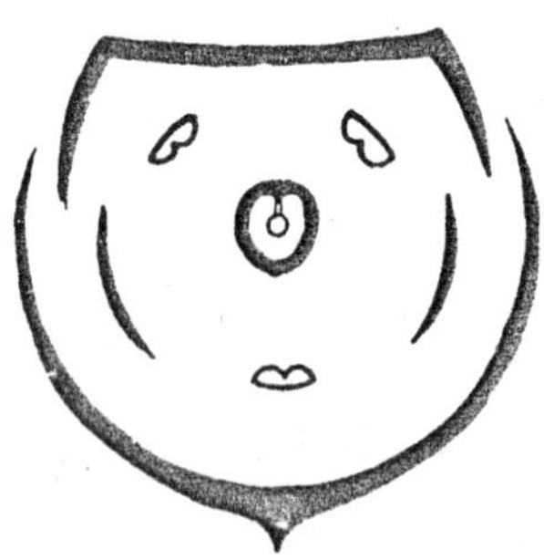

Fig. 8.14. Floral diagram of Gramineae

18

- Used as food:
  Date: *Phoenix dactylifera*
  Coconut: *Cocos nucifera*
- Used in industry:
  Wax: on the leaf
  Rattan: stem
  Fiber: veins in leaves

19

There are many members of Gramineae that are cultivated for their grains:
The common oat: *Avena sativa*
Wheat: *Triticum sativum*
Maize: *Zea mays*
Barley: *Hordeum vulgare*
Rice: *Oryza sativa*
Rye: *Secale cereale*
Sugarcane (*Saccharum officinale*) contains sugar in the pith of the stem.
Other Gramineae are used in industry:
Ammophila: *Ammophila arenaria*
Bamboo: *Bambusa vulgaris*
Some are ornamental:
Bamboo: *Bambusa metake*
Gynerium: *Gynerium argenteum*
Finally, Gramineae form the mainstay of grasslands and lawns.
Landscape gardeners use mixtures that have been studied depending upon the requirement, the soil and the climate of the future lawn, which, among others, include:
Agrostis (*Agrostis tenuis*)
Sheeps' fesque: (*Festuca ovina*)
Timothy grass (*Phleum pratense*)
Kentucky bluegrass (*Poa protensis*)
Rye grass (*Lolium italicum*)

Chapter 9

Plant Geography and Ecology

Plant geography and ecology deal with the study of plants in their natural environment.
All plants are subjected to the effect of various factors. These include:
— water,
— light,
— temperature,
— wind, and
— soil.
In response to each of these factors, plants may react differently:
a) The natural conditions may be satisfactory or they may be such that a plant may have to adapt to them. In both cases the growth of such species is possible.
b) The natural conditions may be too difficult. In this case the growth of a particular species is impossible.

1
Study of Individual Environmental Factors and Adaptation of Plants to each of them

1.1 Water

Water is "reasonably" essential for a plant (*see chapter on "Physiology"*).
The natural environment may:
— either have no water: dry environment,
— or have an excess of water: humid environment.

1.1.1 Dry environment

As in sandy areas: deserts, dunes.
As on rocky surfaces: some mountains.
Where water is frozen: unusable water.
Where the water contains "other matter": salt water.

Plants react either by:
a) Storing water: stem of the Cereus, the sedum leaf.
b) Increased water absorption: development of a deep root system (towards available water by increasing the volume of explored soil).
c) By reducing evaporation of water:

— smaller size of leaves: conifers, heather, genista;

— leathery leaves: laurel, evergreen oak;

— movement of leaves . . .;

— and even warfism of the plant.

d) By adopting a rapid growth cycle during a few months when external conditions are most favourable. Sometimes, the leaves fall during bad weather: deciduous species.
In France, bad weather (dry for plants) is in winter, when plants shed their leaves.
If kept in greenhouses they continue to grow.
It may be noted that in this case, it is not only dry weather that is responsible.

1.1.2 Humid environment

As in the case of aquatic plants: the water lily, watercress.
There is no transpiration and, therefore, there are no stomata.
The water "bears" the plant. There is, therefore, no supportive tissue.
The plant must, however, "float". For this there are many air cavities.

1.2 Temperature

For every plant there are minimum and maximum temperatures for growth, the limits being different depending on its region of origin.
At temperatures lower than 0°C unprotected herbaceous parts are frozen.
High temperatures result in the same symptoms as during very dry weather.
Plants adapt to conditions of temperature in the same way as they do to resist drought (*see Chapter 1, §1.1.9*).

1.3 Light

Light is essential for plants to ensure photosynthesis.

Apart from the fact that there are plants that prefer shade and those that prefer sunlight (*see "Physiology"*), two plants of the same species, one kept in the shade and another in sunlight, will not develop in the same way.
In shady conditions plants tend to fade.
In bright sunlight plants adapt as they do in very dry weather.

1.4 Wind

When conditions are extremely difficult, plants, to different extents, resist violent winds by dwarfism.
In exceptional cases growth is totally suppressed.

1.5 Soil

Plants react from two main points of view.

1.5.1 From the physical aspect, soils may be, to varying extents: humic, sandy, calcareous, clayey (*see course material on Agrology*).

1.5.2 From the chemical point of view, soils may be, to varying extents: acid or basic (*see course material on Agrology*).
In both cases it is clear that the possibility for growth of a particular plant depends on these conditions.
For example, we have:

— undergrowth in humic and acid soils,
— vine rootstock that is resistant in calcareous and basic soils,
— tamarix that grows along the coast in somewhat saline soils and conditions.

2 Geographical Distribution of Plants

Depending on the geographical region under consideration the various factors that we have studied are combined to various extents.

2.1 World flora

The world 'flora' generally provides a useful indication to the horticulturist as it indicates, depending on the origin of a plant, its conditions of natural growth.

2.1.1 Plants from equatorial regions

The climate is hot and humid and the rainfall is heavy.
The relative humidity is high.
Growth is intense and the undergrowth is dark.
Flora: ferns, *Orchidaceae*.

2.1.2 Plants from tropical regions

Tropical regions are characterised by alternate periods of rain.
— In winter, the weather is dry,
— Summer is the rainy season.
The vegetation follows this cycle of variation.
Flora: We note the appearance of deciduous plants.

2.1.3 Plants from subtropical regions

In these regions, where we observe high and also very low temperatures, there are long periods of dry weather.
The vegetation is the same as in "deserts".
Flora: succulent plants.
In desert areas there are water points that have enabled the development of oases.
Flora: palms, orange trees (*Citrus sinensis*), fig trees (*Ficus carica*), etc.

2.1.4 Plants from temperate regions

2.1.4.1 A mainly dry climate is the Mediterranean climate that we have in France along the Riviera.

Flora: laurel (*Laurus nobilis*), holm oak (*Quercus ilex*), olive (*Olea europea*).

2.1.4.2 A predominantly humid climate is the actual climate of France (*see* §*2.2*).

2.1.5 Plants from polar regions

These plants resist cold and dry weather by dwarfism. Eventually the last traces of vegetation are represented by lichens and mosses.

2.2 *The flora of France*

France is situated in the temperate, humid zone. Nevertheless, the vegetation as a whole is subjected to other influences.

2.2.1 Maritime influence

The western region of France has a heavy rainfall with moderate temperature variations between summer and winter.

Flora: English oak (*Quercus robur*), beech (*Fagus sylvatica*), cluster pine (*Pinus pinaster*) and, in some more humid areas, poplars, white willow (*Salix alba*), etc.

2.2.2 Continental influence

This is experienced in the eastern region of France, where there are considerable variations in temperature between the summer and winter.

Flora: mainly conifers: pine (*Pinus sylvestris*), fir, spruce (*Picea excelsa*).

2.2.3 Influence of the mountains

This is felt in the central region especially in the Vosges, the Jura, the Alps, and the Pyrenees, where we may usefully study the growth of vegetation with an increase in altitude because, simultaneously, there is a greater temperature variation and more precipitation.

These characteristics are the same as noted in respect of latitude when we move towards the polar regions.

Flora: This depends, therefore, on the altitude:

a) In the plain, there are normal plantations.

b) Next, we observe forests that persist till 1500–2000 metres. First, there are oak trees (*Quercus robur*), and then beech (*Fagus sylvatica*). Next, there are white birch trees (*Betula verrucosa*) and then conifers.

c) After this there are dwarf trees.

d) Finally, we encounter lichens.

2.2.4 Regional influence of the dry temperate zone: as along the Mediterranean coast, where the following may be grown:

Flora: orange (*Citrus sinensis*), lemon (*Citrus limon*), mimosa (*Acacia dealbata*), cork oak (*Quercus suber*), etc.[1]

1

Ecological knowledge is essential for horticulturists because it enables them to adapt their cultural techniques to the actual requirements of the plant.
For example: a plant may be a short-day plant or need a low night temperature, an acid soil, etc.

Bibliography

Le bon jardinier (The Good Gardener), 2 vols., new ed. 1987. Flammarion.

Coutanceau. *Encyclopédie des jardins* (Garden Encylopaedia). 1973. Larousse.

La vie étrange des fleurs de Jaeger (Jaegar's Remarkable Life of Flowers). 1959. Horizons de France.

Les manipulations de botanique de Genèves (Botanical Manipulations by Genèves). 1962. Dunod.

Sciences naturelles de Chadefaud et Régnier (Chadefaud and Regnier's Natural Sciences). 1961. Delagrave.

Moreau. La vie des plantes de Guillaumin (Guillaumin's Plant Life). 1961. Larousse.

Valdeyron. *Génétique et amélioration des plantes* (Plant Genetics and Improvement). 1961. J.-B. Baillière.

P. Jean-Prost. *La botanique agricole, applications agricoles et horticoles* (Agricultural Botany, Agricultural and Horticultural Applications). 2 vols., 5th ed. 1980.

R. Bossard and P. Cuisance. *Botanique et techniques horticoles* (Botany and Horticultural Techniques). 1977. J.B. Baillière.

H. Vidalie. *La culture* in vitro *et ses applications horticoles* (In Vitro Culture and Its Horticultural Applications). 3rd ed. 1989. Technique et Documentation (Lavoisier).